故宫藏影

西洋镜里的皇家建筑

The Photographic Collection of the Palace Museum
Imperial Buildings through Western Camera

单霁翔 主编

序言

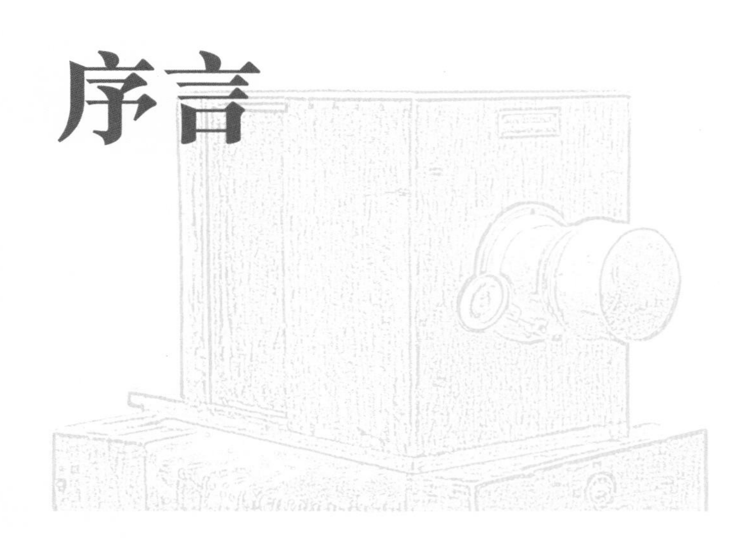

1822 年，法国人尼埃普斯（Joseph Nicéphore Nièpce, 1765~1833 年）拍摄了人类历史上第一张照片——《桌上的物品》。17 年后，1839 年 8 月 19 日，法国科学院向全世界正式公布了由达盖尔（Louis Jacques Mandé Daguerre，1782~1851 年）发明的“银版摄影法”，标志着摄影术这一对人类历史产生深远影响的伟大发明的诞生。

其实，中国人与摄影术的渊源并不比西方人晚。早在两千多年前，中国的墨子就发现了摄影术必备的重要光学原理——“小孔成像”，并记录在《墨经》中流传下来。就在银版摄影法公布的四年后，法国人于勒·埃迪尔（Jules Itier, 1802~1877 年）在 1843 年携带整套达盖尔摄影器材来到了中国，拍摄了广州、澳门一带的风景照以及当时少数中国人的人像照，这是保存至今的最早的中国照片。同年，一位名叫邹伯奇的中国人独立制作出了属于中国人的第一架相机——摄影器，并摄得“平远山水”一幅。可以说，于勒·埃迪尔与邹伯奇是揭开中国近代摄影史序幕的两位开创者。

故宫博物院现存清末以来各种基质的照片近四万张（件），拍摄时间最早可以上溯至19世纪六十年代，所摄内容以清末民国人物、宫廷建筑、文物藏品为主。在为数众多的影像收藏中，众所周知的拍摄于1903年的慈禧太后系列照片，总量在700张以上；包括紫禁城、西苑三海、西郊园林在内的大量宫殿园林照片，在很大程度上指导着今日对现存古建筑的保护与利用；19世纪八九十年代清宫曾拍摄过一批参与筹建北洋海军的官吏组照，众多影响中国近代史的人物影像得以保存；反映溥仪退位后“小朝廷”生活的历史照片及其日后寓居天津的生活掠影；还有民国时期在政治、文化、实业、教育、军事、外交等方面的知名人士等等，这些均是故宫博物院在影像收藏方面的特色种类。

《故宫藏影》是在故宫博物院收藏的近四万张照片中遴选出的精品，分为皇家建筑、宫廷人物与洋务实业三卷，其中相当一批照片是首次公布于世。在这套书中，编者尽力将照片的拍摄地点、或精确或近似的拍摄时间标明出来，并相应地选配一些说明文字，目的就是要将每一张照片的历史信息尽可能完整地展现给读者，使这些照片的价值得到充分彰显。

在皇家建筑一卷中，收录了紫禁城、西苑三海、西郊园林、皇家陵寝等与宫廷有关的建筑照片400余帧，其中尤以大量的紫禁城内照片最具特色。在这批照片中，从1900年外国摄影师拍摄的最早一批紫禁城内部照片，到20世纪四五十年代作为开放的博物馆而留下的馆室外景照，时间跨度长达半个世纪之久，而这半个世纪正是紫禁城六百年历史上最为起伏动荡的一段时间。这期间，既有外敌入侵，共和蒙难，国宝播迁的艰辛，又有建立博物院，大师云集，宣传文化精髓的气概；既有军阀禁锢，日伪统治的屈辱，又有奋力革命，改朝换代的豪迈……可以说，紫禁城半个世纪的风霜雨雪，与中华民族奋勇不屈、力争自由的革命历史紧密相连。这半个世纪的缩影，凝聚在一帧帧灰白照片中，这是紫禁城的记忆遗产，更是民族的历史财富。

在宫廷人物一卷中，我们可以借助照片，与近代中国历史上的那些显赫人物们进行一次时间的邂逅。从慈禧太后到逊清皇室，从宗室王公到封疆大吏，近代历史上能与宫廷沾边的人物大多被收入卷中。当然，除了作为统治阶级出现的著名人物外，往日宫廷中那些终日忍受屈辱，将自己的青春与前途全部埋葬在封建宫廷中的太监宫女们也是这一分卷着力表现的内容。他们的生活状态是什么样子的？非人心理压力下的繁重工作，又还要承受怎样的精神折磨？这些问题都会以照片的形式向读者公开。虽然我们无从知晓那些将镜头对准宫女太监的摄影师们所抱持的立场与心态，我们甚至不知道照片中人物的名字，但是，从一张张或麻木或憧憬，或平和或新奇的面孔中读出的信息是如此的真实，为我们今天反思历史、控诉君主时代的宫监制度提供了绝佳的材料。

在洋务实业一卷中，我们从照片中得以了解前人在将中国引向近代化过程中所经历的那些艰苦卓绝的探索，虽然筚路蓝缕，誓与列强争锋！19 世纪六十年代以来，中国掀起了一股以自强、求富为目的的自上而下的洋务运动。先是官办学堂、工厂在各地兴起，继以民营企业、民族资本一开中国近代资本主义实业的先河。近代洋务运动虽然是以维护清政府统治为目的，但客观上对中国近代化进程产生的影响是深远的：洋务运动中以国家力量开办的大型经济实体，不论军需民用，均在一定程度上改变着这个国家的面貌，在“中体西用”思想指导下，中国近代化的工业基础得以初步打造；还有那些毕业于新式学堂或公派留洋的学生们，他们中的很多人活跃在清末至民国的各种国家舞台上，扮演着推动中国历史车轮不断前进的领航者的角色。以上历史背景下的珍贵镜头，均被收录于洋务实业分卷中。

《故宫藏影》浓缩了中国近代的宫廷史、建筑史、经济史、军事史……当然，还应该包括摄影技术史。我们希望广大读者通过本书审视浓缩的历史时，能够引起进一步的思考：回眸百年历史，我们一路从何走来？因为，只有了解过去，才可以立足现在，终将面向未来。

单霁翔

故宫博物院院长

二〇一四年八月于紫禁城

Preface

In 1822, Joseph Nicéphore Nièpce, a French inventor, made the earliest photograph. 17 years later, on 19th August, French Academy of Science announced the invention of Daguerreotype, which produced by Louis-Jacques-Mandé Daguerre and Joseph Nicéphore Nièpce, to public. The announcement have marked that the photography was born at the year of 1839.

In terms of the study of photography, it is worthy to point that Chinese people have had a further understanding of photography no later than the Western world. For example, Mozi, a Mohist philosopher who was actively in ancient China, mentioned the effect of an inverted image forming through a pinhole 2000 years ago. This discovery, also recognized as the "pinhole camera", recorded in his book *Mozi*.

Four years after the announcement of the invention of Daguerreotype, Jules Itier, a French amateur daguerreotypist, went to China with his camera. He took a number of daguerreotypes of scenery of Guangzhou, Macao and some Chinese people. Meanwhile, a Chinese guy named Zou Boqi independently designed a "camera" by himself, called photographic machine. He later made a picture of "*Pin Yuan Shan Shui*" from his camera. Generally speaking, Jules Itier and Zou Boqi were the pioneers of introducing photography to China in modern Chinese history.

The photographic collection of the Palace Museum primarily comprises more than ten thousand pictures that the earliest photograph could be stemmed from 1860s. This collection features with Chinese people and imperial houses, drawing the timeline from late Qing dynasty to the Republic of China. Therefore, the collection could be divided into 5 parts. The first part is the famous series of Empress Dowager Cixi, who took these pictures during 1903; the second part is the imperial gardens (the Forbidden City and the Sea Palace), which to some extend guides the preservation of the traditional buildings nowadays; the third part is that a number of pictures of important courtiers from 1880s to 1890s, who was photographed by the Imperial Court of late Qing dynasty; the fourth part is the collection of the Last Emperor, Puyi, featuring with the daily life after his abdication and the time he stayed in Tianjin; the last part is about the political and cultural life in the period of Republic of China. In summary, it could be said that the Palace Museum has the most unique photographic collection, comparing the collections with the other institutes.

The photographs from these book series, *The Photographic Collection of the Palace Museum*, are selected from the thousands of pictures of the Palace Museum. There are three volumes of the book: Volume I, Imperial Buildings; Volume II, Imperial People; Volume III, Technological Modernization. Most of the photos from the book are the first time to be publicly published. In order to exhibit the detailed historical information to readers, the editor of the book tries to mark the time and location of every print; and some selected photos are captioned with a introduction.

In the volume of Imperial Buildings, it comprises 400 photos of the Forbidden City, the Sea Palace, imperial gardens of western suburb, imperial temples and altars, and imperial mausoleums. From all these photos, the most significant exhibition prints of this volume are the Forbidden City. The volume provides a series of photographs, which are taken from the inside and outside of the Forbidden City, from the year of 1900 to the year when the Forbidden City had reopened as the Palace Museum. The collection exhibits a glamorous historical document of the Forbidden City through almost five decades. During these decades, the Forbidden City had been through great changes: from the Qing dynasty to the Republic of China, the Second Sino-Japanese War and the Forbidden City reopened as the Palace Museum, etc. All these changes have been connected closely to Chinese people, and this historical photographic collection is the most valuable treasures in Chinese history.

In the volume of Imperial People, it comprises the pictures of Empress Dowager Cixi, the Last Emperor, members of imperial family and ministers of late Qing dynasty. Besides that, this volume also features with the eunuchs and imperial maids in the imperial court. From those pictures, we neither know the names nor the feelings of the eunuchs and imperial maids, however, their daily life and how they service the imperial family could be exhibited vividly through the photos.

In the volume of Technological Modernization, it comprises the photographs of the Self-Strengthening Movement. The Self-Strengthening Movement was a period of institutional reforms that started from 1860s. The movement firstly began with the adoption of Western scientific technology and training of technical and diplomatic personnel through the establishment of a diplomatic office and a college. Later it had moved to the phase of commerce, industry and agriculture. Numerous enterprises that were operated by merchants had been built in this period, and it brought new development of industries such as shipping, railways, mining, and telegraphy, which were rather new ventures for the Qing government. From the photographic collection of Technological Modernization, it shows the changes of modern Chinese industries; the students graduated from colleges and the overseas students granted by the government. All these people play an important role in the modern Chinese history.

Consequently, *The Photographic Collection of the Palace Museum* demonstrates a specialized modern Chinese historical document such as imperial history, architectural history, economic history, military history, and of course the history of photography. Therefore, we hope our readers could think of a question while viewing these historical photographs: are there any factors keep driving the human beings from the past to the future? Maybe we could find our answers through this historical photographic document.

Shan Jixiang
Director of the Palace Museum
August 2014

目录

序言 …… 6
上篇　从紫禁城到博物院 …… 16
紫禁城外朝中路（含外三门） …… 25
紫禁城外朝东路、西路 …… 73
紫禁城内廷中路 …… 87
紫禁城内廷东路、西路 …… 151
紫禁城内廷外东路 …… 179
紫禁城内廷外西路 …… 197
下篇　皇家苑囿与陵寝 …… 216
景山 …… 225
西苑三海 …… 237
圆明园 …… 263
清漪园（颐和园） …… 269
静明园 …… 297
静宜园 …… 309
避暑山庄 …… 317
农事试验场 …… 327
黑龙潭龙王庙 …… 335
乐净山斋 …… 339
醇亲王园寝 …… 343
清西陵 …… 353
图版索引 …… 386
编后记 …… 401

Contents

Preface 10
From the Imperial Palace to the Palace Museum 20
The Middle Section of Outer Court of the Forbidden City 25
The Eastern and Western Section of Outer Court of the Forbidden City 73
The Middle Section of Inner Court of the Forbidden City 87
The Eastern and Western Section of Inner Court of the Forbidden City 151
The Outer Eastern Section of Inner Court of the Forbidden City 179
The Outer Western Section of Inner Court of the Forbidden City 197
Imperial Gardens and Mausoleums 220
Jingshan 225
Xiyuan Sanhai (Sea Palace) 237
Yuanming Yuan (Old Summer Palace) 263
Qingyi Yuan (Summer Palace) 269
Jingming Yuan 297
Jingyi Yuan 309
The Summer Resort 317
Experimental Farm 327
The Dragon Temple of Heilong Spring 335
Lejing Shanzhai 339
Tomb of Prince Chun 343
West Mausoleum of Qing dynasty 353
Index 393
Prologue 404

上篇
从紫禁城到博物院

北京紫禁城自明永乐十八年(1420年)建成,到清朝最后一个皇帝溥仪退位、清朝灭亡,经历了491年的历史。其建筑以南北中轴线上的乾清门广场为界分为南北两部分,南为外朝理政之地,北为内廷居住之所。1911年辛亥革命后,根据民国政府“大清皇帝辞位之后,暂居宫禁,日后移颐和园”的优待条件,逊帝溥仪仍居内廷。外朝地区为国民政府所辖。时以乾清门广场为界,砌筑一道高墙,将外朝、内廷分隔开来。1914年成立的古物陈列所,利用外朝中路主体建筑中的太和、中和、保和三大殿,东西两翼的文华、武英等殿为展示历朝珍品文物的展室,向社会开放,宫禁大门始开启。外朝的开放,使得紫禁城恢宏的建筑向世人展示，遂有机会留下此时大量的照片。1916年袁世凯称帝，欲在太和殿登基，命将外朝建筑所有满汉文匾一律改写为汉文匾，故今日外朝建筑匾文均为汉文。太和殿汉文匾之照片记录了这段历史。1924年，冯玉祥发动“北京政变”，将溥仪逐出宫禁。成立“清室善后委员会”，清点查验清宫文物。1925年故宫博物院成立，以内廷为院址，利用内廷主要建筑作为博物院的展室，对外开放。至此，紫禁城全面开放。神武门城墙上有块“故宫博物院”匾，说明此片所摄已是1925年故宫博物院成立后的紫禁城。此时照片中还清晰可见神武门外北上门及清代景山官学的所在地——北上门两旁的长房。而东华门外所挂牌子上标明的“国立北平故宫博物院由神武门出入”，说明外朝、内廷依然分隔，凡到故宫博物院的参观者只好由神武门进出。而有“故宫博物院”匾的午门全景照片，记录的则是1948年古物陈列所并入故宫博物院，实现了故宫完整统一管理的历史。

外朝建筑是紫禁城建筑的主体，建筑在三层汉白玉台基之上的三大殿位居紫禁城中心，其建筑雄伟，气势恢宏。与四隅崇楼、左体仁阁、右弘义阁、前后九座宫门以及周围廊庑，共同构成了占地约八万平方米的紫禁城内最大的庭院。太和殿是明清两朝帝王举行盛大典礼的场所，凡皇帝登基大典、大婚、册立皇后、命将出征，以及元旦（春节）、冬至、万寿（皇帝生日）三大节，皇帝都要在此接受朝贺，并在此赐宴。这座清康熙三十六年（1697年）重新建成的太和殿，是中国现存体量最大、建筑等级最高的木结构建筑，它不仅经历了三百年的风风雨雨，见证了帝王的威严、时代的盛衰，经历了外来侵略的屈辱，留下了袁世凯做了83天洪宪皇帝的遗迹；也记录了1945年10月10日，在全中国人民走过八年抗战的艰难历程，故宫博物院建院20周年之日，太和殿前举行的隆重的受降仪式，太和殿及数万群众见证了这一辉煌的历史时刻。

享有民国政府“岁用四百万银元”拨款，“侍卫人等，照常留用”之待遇，逊帝溥仪在紫禁城内廷又居住了13年。内廷建筑拍摄的时间多为1911年至1924年期间，这些照片记录了这段鲜为人知的历史。

内廷建筑照片涉及：内廷中心主体建筑乾清宫、交泰殿、坤宁宫，俗称后三宫，是为帝后寝宫；后三宫北部的御花园；众妃嫔居住的东西六宫，以及从清代雍正年间成为皇帝寝宫的养心殿和雍正年间建成的斋宫，以上建筑都位于内廷的核心区域。在此之外，还有西部供皇太后居住的寿安宫，慈宁宫、慈宁宫花园以及乾隆初年建成的建福宫花园；东部有乾隆年间改扩建的宫殿宁寿宫、宁寿宫花园等，基本涵盖了内廷的主要角落。

乾清宫是明代皇帝的寝宫，自永乐皇帝朱棣至崇祯皇帝朱由检，共有14位皇帝在此居住过。清初，乾清宫虽然在名义上还是皇帝的寝宫，但是顺治、康熙两位皇帝继位之初都没有在乾清宫居住。顺治皇帝福临在顺治十三年（1656年）才移居乾清宫。康熙皇帝玄烨自康熙八年（1669年）才居住乾清宫，直至六十一年在此居住了50多年。康熙皇帝死后，梓宫奉安在乾清宫，雍正皇帝在养心殿守丧，此后养心殿即为清代皇帝的寝宫。从明到清，在乾清宫发生过太多的故事。老照片中的乾清宫，则为清代嘉庆三年（1798年）重建。

坤宁宫是内廷的中宫，明代皇后在此居住，主内廷统摄六宫。清顺治十二年（1655年），在重修后三宫时，仿盛京皇宫的清宁宫形制，对坤宁宫做了较大的改建，将原明间正门移到东次间，改为板门，其他几间撤去棱花隔扇门改为直棂吊窗。室内东次间与西三间改为满族萨满教祭神的场所，内设神龛、供案、置办祭品的煮肉大锅等，此后坤宁宫就作为满族祭祀萨满神的主要场所。东两间隔出后作为暖阁，供人居住，自康熙皇帝玄烨在此大婚后，这里就成为清代皇帝大婚的洞房。由于康熙以后的五位皇帝在继位前已婚娶，所以直至同治皇帝大婚，坤宁宫才再次迎来喜庆气氛。光绪十五年（1889年）皇帝大婚，也是在这里举行。而照片中洞房的一切陈设，应为紫禁城里的最后婚礼——溥仪结婚的场景。

养心殿是一处独立的院落，位于乾清宫西侧，明嘉靖十六年（1537年）建造，清初顺治皇帝逝于此地。自雍正皇帝移驻养心殿一直到溥仪出宫，清代有八位皇帝曾在这里居住。1911年辛亥革命后，退位的溥仪继续在养心殿居住。由于皇帝居住的原因，这组建筑经不断地改造、添建，成为集召见臣工、处理政务、皇帝读书及居住为一体的多功能建筑群。今天我们从百年照片中，看到了雍正年间添建的养心殿抱厦，乾隆年间添建的养心门外值房，曾经是两宫皇太后垂帘听政的养心殿东暖阁，慈禧、慈安居住过的燕禧堂、体顺堂，这些或尚存原状，或已变化，而留下更多的是溥仪居住时闲散安逸的生活场景。

清末，电灯、电话这些舶来品就已经进入清宫，皇帝的寝宫养心殿、处理政务的乾清宫以及后妃等居住的各处宫殿均安装玻璃吊灯，与清代建筑及室内陈设形成了极大的反差，也融入了时代的色彩。

御花园内的竹篱笆墙、鹿囿，养性斋、绛雪轩前高高的遮阳棚架，今天已无所见。照片中慈禧太后曾经居住过的宁寿宫里的乐寿堂，也已无人问津，竹帘低垂，景象凄凉。而在上万张老照片中，几张珍贵的建福宫花园烧毁前的照片，让我们看到了昔日备受乾隆皇帝喜爱的这座花园的富丽辉煌。

从养心殿翠绿旺盛的盆景到摆满庭院的秋菊，从大大小小的鱼缸到凌乱无序的卧室，在年复一年、日复一日的享乐之中，在建福宫花园的熊熊大火之中，溥仪的紫禁城生活走到了尽头。

周苏琴

故宫博物院研究馆员

From the Imperial Palace to the Palace Museum

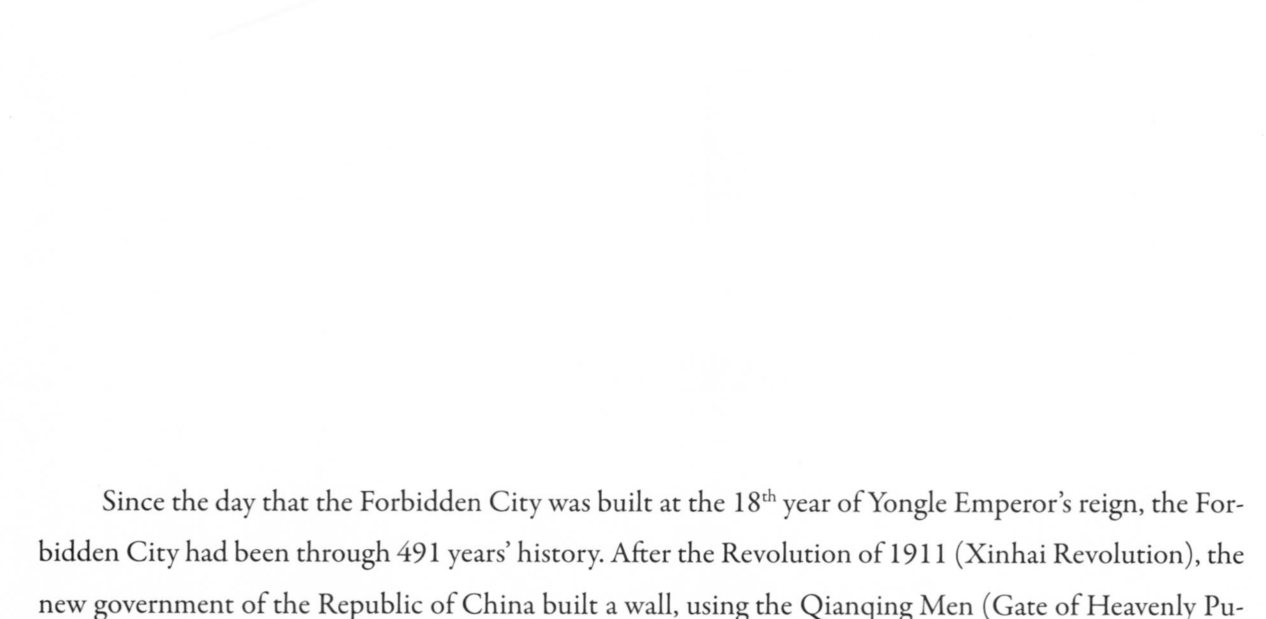

Since the day that the Forbidden City was built at the 18th year of Yongle Emperor's reign, the Forbidden City had been through 491 years' history. After the Revolution of 1911 (Xinhai Revolution), the new government of the Republic of China built a wall, using the Qianqing Men (Gate of Heavenly Purity) as the border, to separate Inner Court and Outer Court. The Inner Court, according to the Favorable Treatment, was provided as the private area of residence to Puyi, who later had to move to the Summer Palace. The Outer Court was governed by the Republic of China. In 1916, Yuan Shikai, famous for his influence during the late Qing dynasty and the early time of the Republic of China, claimed the throne and planned to be enthroned at the Taihe Dian (Hall of Supreme Harmony), so that he ordered people to replace the tablets engraved with Manchu and Chinese with the tablets only engraved with Chinese. It is the reason why nowadays people could only see the tablets engraved with Chinese at the Outer Court. In 1924, warlord Feng Yuxiang launched "Beijing Incident", and Puyi was expelled from the Forbidden City. Feng established the "Committee for the Aftermath of the Qing Royal Family" in order to organize a committee to in charge of the management of the Forbidden City. Later in 1925, the Forbidden City reopened as the "Palace Museum" to public, by using the Inner Court as the exhibition hall at the same time.

Basing on the construction of the Forbidden City, the whole palace could be divided into the Outer Court and Inner Court, by the square of Qianqing Men (Gate of Heavenly Purity) standing right on the south-north axis. Therefore, this section of photographs will be divided into two parts based on construction: Outer Court and Inner Court. The Outer Court are the main body of the Forbidden City. There are three magnificent halls stand on top of a three-tiered white marble terrace of the Outer Court. These are Taihe Dian (Hall of Supreme Harmony), Zhonghe Dian (Hall of Central Harmony), and Baohe Dian (Hall of Preserving Harmony). Together these three halls constitute the heart and the largest court of the Outer Court of the Forbidden City, measuring approximately 80,000 sq. meters. Besides, Taihe Dian, one of the three halls, is the major hall and it had held imperial ceremonies since Ming and Qing dynasties, such as coronations, investitures and imperial weddings. Furthermore, Taihe Dian was originally built by the Ming dynasty and destroyed seven times by fire and lastly rebuilt at the 36th year of Kangxi Emperor's reign (1697).

The Outer Court was opened as the "Institute of Antique Exhibition" to the public in 1914, by using the Taihe Dian, Zhonghe Dian and Baohe Dian, Wenhua Dian (Hall of Lofty Literary) and Wuying Dian (Hall of Military Prowess) as the exhibition halls. Since then, plenty of the photographs of the Out Court were taken and left to nowadays.

The Inner Court is separated from the Outer Court by Qianqing Gong, and it was the home of emperors and his family. In the Qing dynasty, emperors lived and worked almost exclusively in the Inner Court, leaving the Outer Court used only for ceremonial purposes. There are other three halls constitute the center of the Inner Court; these are Qianqing Gong (Palace of Heavenly Purity), Jiaotai Dian (Hall of Union) and Kunning Gong (Palace of Earthly Tranquility). Together these three halls were the home apartments for emperor and empress. Besides, there are Ningshou Gong (Palace of Tranquility and Longevity), Cining Gong (Palace of Compassion and Tranquility), Garden of Cining Gong and Garden of Jianfu Gong (Jianfu Gong, Palace of Establishing Happiness) from the west; Ningshou Gong (Palace of Tranquility and Longevity) and Garden of Ningshou Gong (popularly known as the Qianlong Garden) from the east. Together all these palaces constitute the whole body of the Inner Court.

Qianqing Gong was the residence of emperors of Ming dynasty. 14 emperors of Ming dynasty had been lived there since the Yongle Emperor. During the early years of Qing dynasty, although Qianqing Gong still was the residence of emperor, Shunzhi Emperor and Kangxi Emperor did not move immediately to the Qianqing Gong after their enthronement. Shunzhi Emperor moved to Qianqing Gong after 13 years of his reign (1656); Kangxi Emperor moved to Qianqing Gong after 8 years of his reign (1669) and lived there for more than 50 years. However, when the Yongzheng Emperor ascended to the throne, he did not move to the palace where his father Kangxi Emperor used to live. He and subsequent emperors lived in a smaller palace, Yangxin Dian (Hall of Mental Cultivation).

Kunning Gong is one of the three main buildings of the Inner Court, and it was the residence of the empress, who was in charge of the daily life of entire Inner Court, of Ming dynasty. In Qing dynasty, large portion of the palace were reconstructed, by using the Qingning Gong (Palace of Pure Tranquility) of Shengjing (Shengjing is the co-capital of the Qing dynasty) as architetural model, during the reign of Shunzhi Emperor. Consequently, the main gate of the palace moved to the room of eastern side; the rooms of eastern side and three rooms of west were converted for Shamanist worship. The rooms featured shrines, icons, prayer meats, and a large kitchen where sacrificial meat was prepared. Since then, Kunning Gong was used as the place for offering the sacrifices to the deity of Shaman. Meanwhile, there were two rooms of eastern side retained for use on emperor's wedding night after Kangxi Emperor held his wedding night there.

Yangxin Dian is an independent palace, locates at the west side of Qianqing Gong. The palace was built at the 16th year of Jiajing Emperor's reign (1537). Since Yongzheng Emperor moved in Yangxin

Dian, eight emperors of Qing dynasty had been living here. After several times of redecoration, the palace had become into a multi-functioning buildings for emperors to attend state affairs. During the late Qing dynasty, Empress Dowagers Cixi and Empress Dowager Ci'an attended state affaits at the Dongnuan Ge (the East Warm Chamber) of Yangxin Dian.

The Inner Court did not open to public till 1924. From 1911 to 1924, the Last Emperor, Puyi, lived in the Inner Court. During this period, many imported goods such as electronic lights and telephones were introduced to China and the Inner Court of the Forbidden City. Almost hundreds of photos were taken and recorded this history that the palaces were installed with lights and telephones, and the daily life of the Last Emperor.

All these precious old photographs demonstrate the glorious time of the Forbidden City before the fire burnt down the beautiful gardens and palaces. From this book, we may have the chance to check the old time of all these magnificent buildings of the Forbidden City.

Zhou Suqin

Senior Researcher of the Palace Museum

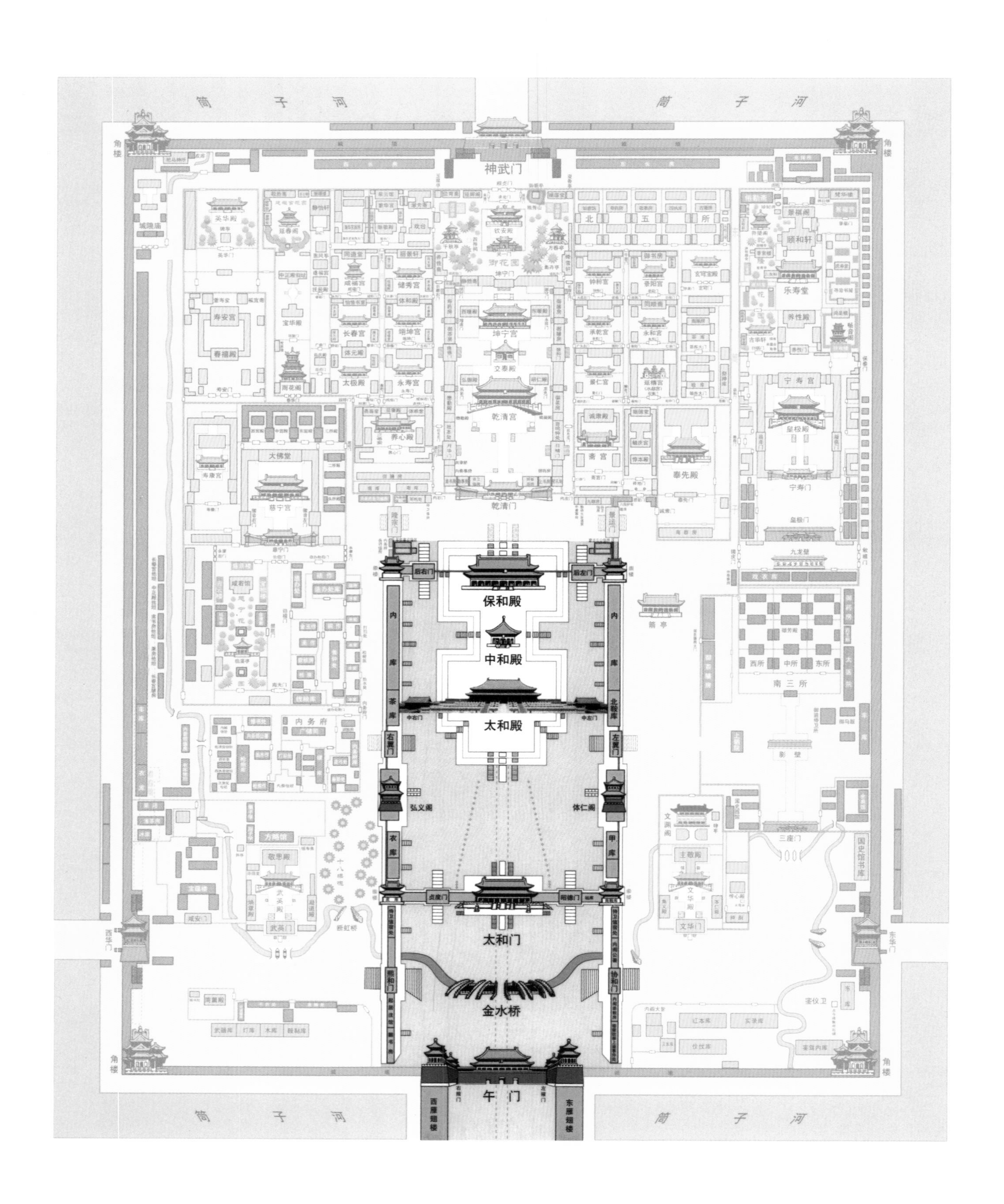

筒子河
筒子河
角楼
角楼
神武门
御花园
坤宁宫
交泰殿
乾清宫
乾清门
养心殿
大佛堂
慈宁宫
寿安宫
雨花阁
宁寿宫
皇极殿
宁寿门
皇极门
九龙壁
奉先殿
斋宫
乐寿堂
颐和轩
北五所
保和殿
后右门
后左门
内库
中和殿
太和殿
中右门
中左门
茶库
北裂库
右翼门
左翼门
弘义阁
体仁阁
衣库
甲库
贞度门
昭德门
太和门
熙和门
协和门
金水桥
午门
西雁翅楼
东雁翅楼
内务府
南三所
西所
中所
东所
箭亭
文渊阁
主敬殿
文华殿
文华门
三座门
国史馆书库
武英殿
武英门
敬思殿
方略馆
十八槐
断虹桥
西华门
东华门
角楼
角楼
筒子河
筒子河

紫禁城外朝中路（含外三门）

The Middle Section of Outer Court of the Forbidden City

紫禁城以乾清门广场为界，
南部统称为“外朝”，北部统称为“内廷”。
外朝地域广阔，建筑大气，
其中尤以太和、中和、保和三大殿最为辉煌壮丽，
处处体现了帝王“九五之尊”的威严与肃穆。
外朝中路的核心是太和殿，这里是国家举行重大朝会的场所，
凡遇皇帝登基、大婚、册封皇后、上皇太后徽号、
命将出征以及三大节（元旦、冬至、万寿）等重要日子，皇帝均御太和殿。
这座重建于康熙年间的中国现存最大的木结构建筑，
三百多年来见证了太多的重要历史时刻。

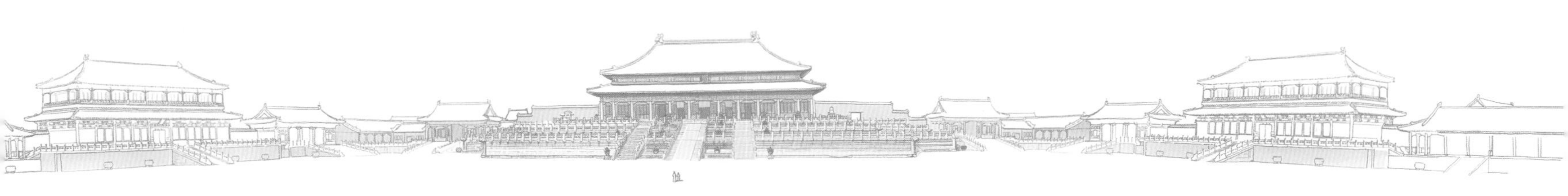

大清门／1901年

Daqing Men (Great Qing Gate), 1901

大清门在明代称大明门，是明清两朝的国门，帝王出巡、回銮的重要门座，门前旷地即著名的“棋盘街”。此帧照片反映了1901年慈禧太后携光绪帝回銮前后，大清门一区修饰一新的场景。

大清门门额／1901年

The Tablet of Daqing Men, 1901

此帧照片中满汉文“大清门”字样赫然在目，1912 年 2 月，清帝退位，民国建立，大清门改名中华门。原大清门门额现藏故宫博物院。

中华门 / 1912~1915年

Gate of China, 1912-1915

1912 年，新换中华门门额由时任内务部次长王治馨题写。照片中处在大清门与天安门之间的是东西千步廊，将翰林院、宗人府、六部等重要部院分隔在中轴线两侧。

中华门 / 1915~1920年

Gate of China, 1915-1920

中华门门额 / 1915~1920年

The Tablet of Gate of China, 1915-1920

1912年，北洋政府换上了黑底金字横写"中华门"门额。1915年，又将横额改换成中轴线统一风格的竖写"中华门"门额。

天安门 / 1900年

Tiananmen (Gate of Heavenly Peace), 1900

天安门初名承天门，在明清两代是国家举行颁诏大典的地方。1900 年 6 月，八国联军进占北京，此帧照片为我们展示了天安门被侵略者炮火击伤后的惨状。

天安门 / 1910~1920年
Tiananmen, 1910-1920

天安门及门前华表 / 1900年
Tiananmen and Huabiao, 1900

此帧照片满汉文匾额清晰可见。据张伯驹回忆，袁世凯窃国期间，袁命国务院参事林长民以汉文重新书写紫禁城外朝匾额。

天安门前金水桥 / 20世纪初

Golden River Bridge in front of Tiananmen, Early 20th Century

天安门前设五路金水桥，称外金水桥，与午门北侧的内金水桥相对应。此帧照片的远景为长安左门，是明清两代殿试宣示“黄榜”的地方，民间有“龙门”之称，今已不存。

天安门前石狮 / 1900 年

The Imperial Guardian Lions in front of Tiananmen, 1900

天安门北侧 / 1900年

North Side of Tiananmen, 1900

端门／1913年3月

Duan Men, March 1913

端门与天安门形制相同，在清代端门城楼是存放帝王出行仪仗的场所。此帧照片拍摄于1913年3月隆裕皇太后丧礼期间，中间三座门洞上装饰着白花。

午门 / 1900年

Wu Men (Meridian Gate), 1900

午门是紫禁城正门，也是紫禁城四座城门中最为高大宏敞的一座。午门的平面呈“凹”字形，沿袭了唐朝大明宫含元殿及宋朝宫殿丹凤门的形制，是从汉代的门阙演变而来。明清两代，午门中门为皇帝专用，帝王出巡、回銮皆从中门进出；皇帝大婚时，皇后乘坐的喜轿可以从中门进宫；通过殿试选拔的状元、榜眼、探花，在宣布殿试结果后可从中门出宫；东侧门供文武官员出入，西侧门供宗室王公出入；两掖门只在举行大型活动时开启。此门亦是战争得胜后，凯旋将士向皇帝行“献俘礼”的场所。

午门 / 1912~1927年

Wu Men, 1912-1927

此帧照片中的午门前，交叉着巨大的五色旗，这是判断照片拍摄年代的重要依据。1912 年 1 月，中华民国尚未诞生，而当时的临时参议院通过了决议，以红、黄、蓝、白、黑五色旗作为临时国旗。直到 1928 年北伐成功，北洋政府倒台，五色旗被“青天白日满地红旗”取代。

午门 / 1948年

Wu Men, 1948

1948 年 3 月，北平古物陈列所并入故宫博物院，故宫博物院第一任院长易培基提出的“完整故宫保管计划”终于实现，午门外正式挂起了“故宫博物院”字样的匾额。

午门西雁翅楼 / 20世纪初
West Yanchi Lou of Wu Men, Early 20th Century

午门东雁翅楼 / 1922年

East Yanchi Lou of Wu Men, 1922

隆裕皇太后在建福宫花园 / 1911年

Empress Dowager Longyu at the Jianfu Gong Garden, 1911

午门北侧（三联照）／1913年3月

North Side of Wu Men, March 1913

1913 年 3 月，隆裕皇太后丧礼期间，北洋政府在紫禁城内召开了隆重的“国民哀悼会”，以悼念隆裕皇太后在清帝退位过程中“力主共和”的“功绩”。照片显示了午门北侧张贴挽联的景象。太和门广场上人头攒动，北洋政府安排了列兵，以壮观瞻。

太和门广场 / 1900年

Taihe Men Square, 1900

太和门明代称奉天门，嘉靖时改称皇极门，明帝在此举行“御门听政”。满族统治者入关后，皇极门改称太和门。光绪十四年（1888年），太和、贞度、昭德三门连同部分廊庑被火焚毁，后重建。

太和门前石亭 / 1900年

A Stone Pavilion of Taihe Men, 1900

太和门前青铜狮与熙和门 / 1900年

A Bronze Lion and Gate of Xihe in front of Taihe Men, 1900

太和门广场上协和门与熙和门东西相对，二门是紫禁城外朝东西两翼进出中路的重要门座。协和门廊庑在明代为实录馆、玉牒馆和起居注馆，清代改作稽察钦奉上谕事件处和内阁诰敕房；熙和门廊庑在明代作为编修《大明会典》的会典馆，清代改为翻书房和起居注馆。

太和门前青铜狮 / 1900年

A Bronze Lion in front of Taihe Men, 1900

太和门前陈设青铜狮一对，为明代铸造。

太和门 / 1913年3月

Taihe Men (Gate of Supreme Harmony), March 1913

此帧照片为隆裕皇太后丧礼期间，众人列队进入太和门致哀的场景。隆裕皇太后的丧礼，北洋政府给予了极高的治丧规格。议定大小官署一律下半旗二十七日，官员左腕围黑纱二十七日，以志哀悼。

太和门 / 1913年3月

Taihe Men, March 1913

太和门内临时搭设的悼棚 / 1913年3月

A Temporary Funeral inside the Taihe Men, March 1913

太和门内景 / 1913年3月

Interior Scene of Taihe Men, March 1913

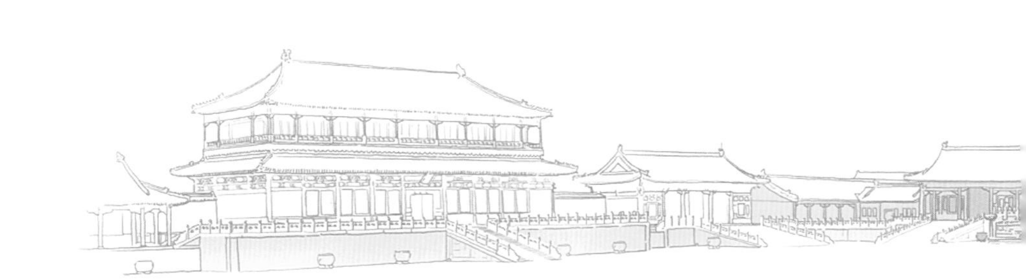

太和殿广场／1900年
Taihe Dian Square, 1900

太和殿广场 / 1918年11月28日

Taihe Dian Square, 28th November 1918

1918 年 11 月 28 日，作为协约国一方的中国为庆祝第一次世界大战的胜利，在太和殿广场举行了盛大的阅兵式。

太和殿广场 / 1918年11月28日

Taihe Dian Square, 28th November 1918

从太和门北侧望太和殿 / 1925~1949年

Taihe Dian, from the angle of north Taihe Men, 1925-1949

太和殿 / 1915~1937年

Taihe Dian (Hall of Supreme Harmony), 1915-1937

太和殿在明代称奉天殿，居于紫禁城中轴线的核心，面阔十一间，重檐庑殿顶，是整个紫禁城规模最大、等级最高的殿宇，处处体现着帝王的“九五之尊”。太和殿是明清两代国家举行重大典礼的场所。凡皇帝登基、大婚、册封皇后、上皇太后徽号、命将出征以及万寿、冬至、元旦“三大节”等，皇帝均御太和殿；清初，科举考试的最高一级“殿试”在这里举行；乾隆五十四年（1789年）殿试改在保和殿后，宣布考试结果的“传胪”典礼仍然安排在太和殿内。

太和殿前丹陛石 / 1915~1937年

Danbi Shi (stone carved with clouds and dragons) in front of Taihe Dian, 1915-1937

太和殿前有宽阔的平台，称为丹墀，俗称月台。月台上陈设着铜龟、铜鹤各一对，是长寿的象征；日晷、嘉量各一座，日晷是古代的计时器，嘉量是古代的量器，象征着时间与空间在皇权的主导下；另外还有铜鼎十八座。殿下为八米多高的三层汉白玉须弥座，周围环以栏杆。栏杆下安有 1142 颗排水用的石雕螭首，每逢雨季，可呈现千龙吐水的奇观。

太和殿前铜仙鹤 / 1900年

A Bronze Crane in front of Taihe Dian, 1900

太和殿前铜龟 / 1900年

A Bronze Turtle in front of Taihe Dian, 1900

太和殿前嘉量 / 1900年

A Jia Liang in front of Taihe Dian, 1900

太和殿前日晷 / 1900年

A Sundial in front of Taihe Dian, 1900

太和殿内宝座与天花藻井 / 1900年

Throne, Caisson and Ceiling of Taihe Dian, 1900

太和殿内景／1900年

Interior Scene of Taihe Dian, 1900

太和殿内的装饰十分豪华。檐下施以密集的斗栱，室内外梁枋上饰以和玺彩画。门窗上部嵌成菱花格纹，下部浮雕云龙图案，接榫处安有镌刻龙纹的镏金铜叶。殿内金砖铺地，明间设宝座，宝座两侧排列六根直径一米的沥粉贴金云龙图案的巨柱，所贴金箔采用深浅两种颜色，使图案突出鲜明。宝座前两侧有四对陈设：宝象、角端、仙鹤和香亭。宝象象征国家的安定和政权的巩固；角端是传说中的吉祥动物；仙鹤象征长寿；香亭寓意江山稳固。宝座上方天花正中安置形若伞盖向上隆起的藻井。藻井正中雕有蟠卧的巨龙，龙头下探，口衔轩辕镜。

宝座上方“建极绥猷”四字为乾隆帝御笔，意为：帝王仰承天命而抚育万民；建立法则而顺应天道。三大殿内全部匾额楹联在袁世凯窃国期间下落不明。

太和殿内金漆宝座与雕龙髹金屏风 / 1900年
Golden Lacquer Throne and Screen Carved with Dragon in Taihe Dian, 1900

太和殿内袁世凯称帝时所用宝座 / 1922年
Throne in Taihe Dian (used by Yuan Shikai, during the period he declared as Chinese Emperor), 1922

袁世凯所用宝座 / 1922年
Throne in Taihe Dian (used by Yuan Shikai, during the period he declared as Chinese Emperor), 1922

1915年，袁世凯窃国称帝期间，对紫禁城外朝三大殿进行了一番修改。首先将三大殿内匾额楹联全部撤下，再将地平上的金漆宝座换成适合袁世凯身材的高背大椅，后来又在殿内装电灯、置佛像，好不热闹。

太和殿内满铺的地毯／1900年

Taihe Dian Covered with Carpet, 1900

太和殿内紫宸台 / 1900年

A Dais in Taihe Dian, 1900

太和殿内沥粉金柱 / 1900年

The Golden Lacquer Pillar in Taihe Dian, 1900

中和殿／1900年

Zhonghe Dian (Hall of Central Harmony), 1900

中和殿在明代称华盖殿、中极殿，明堂九室，形作四方，是三大殿中体量最小的一座。凡遇“三大节”等重大朝会，皇帝先至中和殿升座，内阁、内大臣、礼部、都察院、翰林院、詹事府各堂官及侍卫执事人员行礼毕，皇帝才由中和殿御太和殿。每年的“亲耕礼”前，皇帝要先在中和殿阅视农具与祝版。

中和殿内景 / 1900年

Interior Scene of Zhonghe Dian, 1900

保和殿 / 1900年

Baohe Dian (Hall of Preserving Harmony), 1900

保和殿是皇帝赐宴宗室王公、大臣与外藩使节的场所。顺治、康熙二帝曾一度以保和殿为寝宫，顺治皇帝大婚的洞房也曾设在这里。乾隆五十四年（1789年）以后的“殿试”多在此举行。

保和殿内景／1900年
Interior Scene of Baohe Dian, 1900

从乾清门看保和殿／1900年

Baohe Dian, from the angle of Qianqing Men, 1900

保和殿后云龙大石雕／1900年

The Marble Stone Carved with Clouds and Dragons at the Backyard of Baohe Dian, 1900

保和殿后大石雕在明代开采于北京房山大石窝，重达二百余吨。乾隆年间，为向崇庆皇太后贺寿，乾隆帝命工匠在原石材上重新摹刻图案。

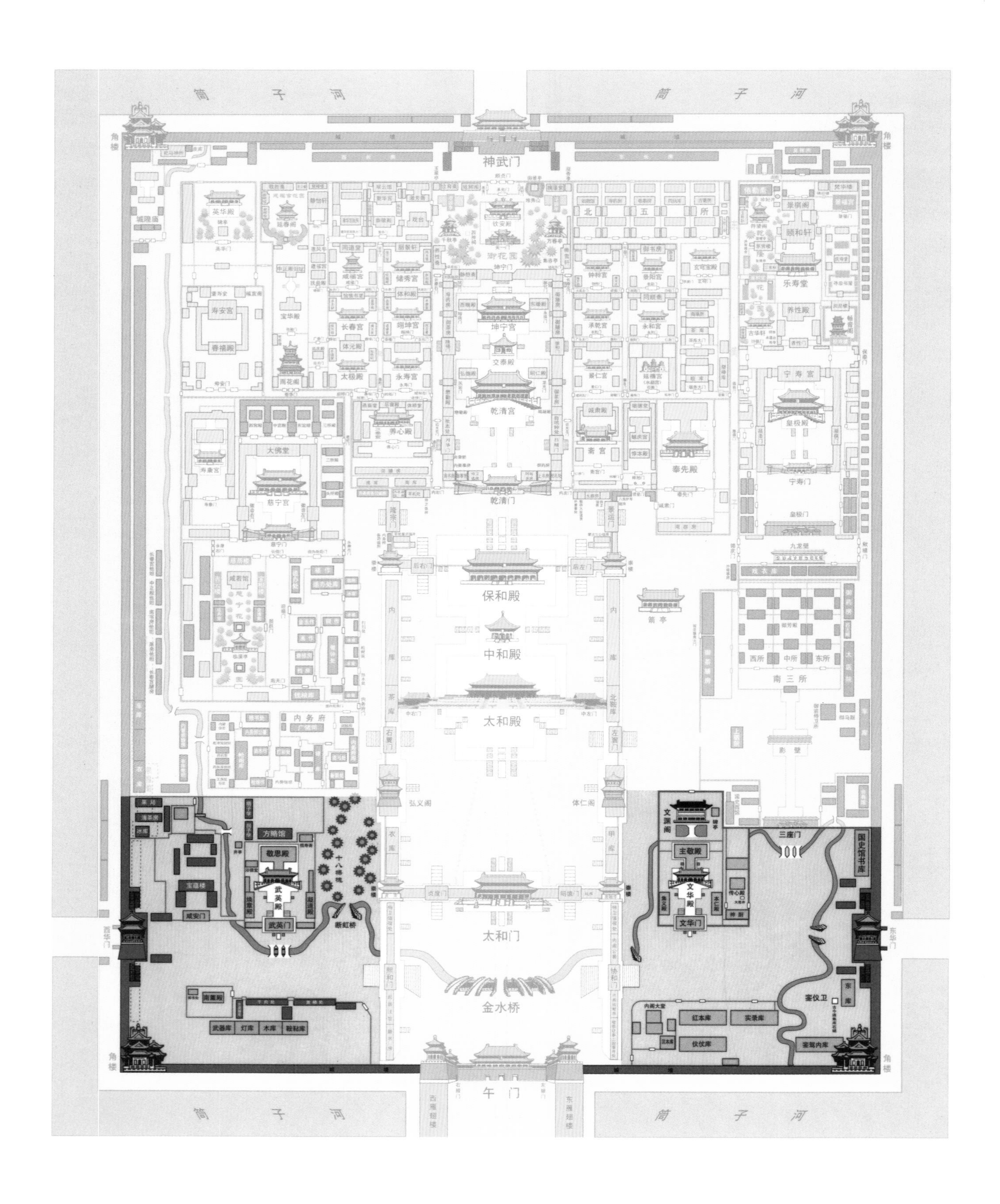

筒子河
筒子河
神武门
角楼
角楼
御花园
北五所
坤宁宫
交泰殿
乾清宫
乾清门
养心殿
慈宁宫
寿安宫
寿康宫
宁寿宫
皇极殿
奉先殿
保和殿
中和殿
太和殿
南三所
武英殿
敬思殿
方略馆
武英门
断虹桥
十八槐
文渊阁
主敬殿
文华殿
文华门
三座门
国史馆书库
弘义阁
体仁阁
太和门
金水桥
西华门
东华门
銮仪卫
东库
内阁大堂
红本库
实录库
仪仗库
銮驾内库
南薰殿
武器库
灯库
木库
鞍鞯库
咸安门
宝蕴楼
内务府
影壁
箭亭
九龙壁
角楼
角楼
午门
筒子河
筒子河

紫禁城外朝东路、西路

The Eastern and Western Section of Outer Court of the Forbidden City

紫禁城外朝东西两路，

“文东武西”分别建有文华殿与武英殿。

文华殿曾是明代“太子视事之所”，清代皇帝在此举行“经筵之礼”，

其后文渊阁是庋藏《四库全书》的藏书楼。

武英殿因康熙年间设立的修书处，

二百年来刊印了大量印装精美的“殿本图书”而名扬海内。

外朝东西两路还有内阁大堂、传心殿、南薰殿、咸安宫等殿宇，

以及实录、红本、仪仗、武器、銮驾等重要的皇家库房。

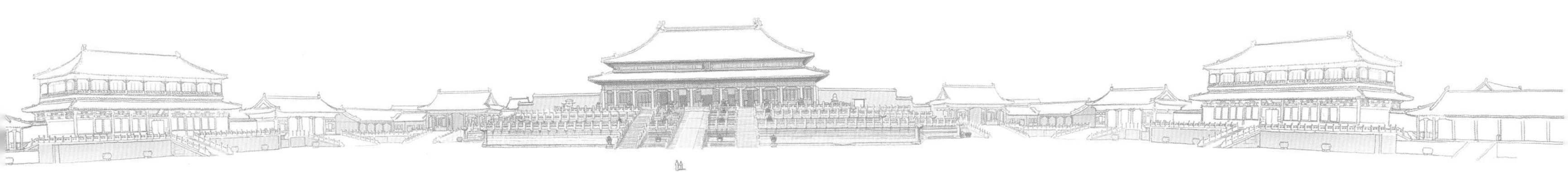

文华殿 / 20世纪50年代
Wenhua Dian (Hall of Literariness), 1950s

文华殿记碑 / 1925~1949年
Stela of Wenhua Dian, 1925-1949

文华殿 / 20世纪50年代
Wenhua Dian (Hall of Literariness), 1950s

文华殿位于协和门以东，与武英殿东西相对。因其位于紫禁城东部，曾一度作为东宫太子的“视事之所”。明清两朝，每岁春秋仲月，都要在文华殿举行“经筵之礼”。清代以大学士、尚书、左都御史、侍郎等人充当经筵讲官，满汉各八人。每年以满汉各二人分讲“经”、“书”，皇帝本人则撰写御论，阐发讲习“四书五经”的心得，礼毕，赐茶赐座。明清两朝殿试阅卷也在文华殿进行。

文渊阁／1900年

Wenyuan Ge (Wenyuan Pavilion), 1900

《四库全书》是乾隆朝最重要的文化工程，一共抄写七部，分藏于热河避暑山庄文津阁、北京圆明园文源阁、北京紫禁城文渊阁、盛京皇宫文溯阁、扬州大观堂文汇阁、镇江金山文宗阁、杭州孤山文澜阁。《四库全书》现存文渊阁本、文津阁本、文溯阁本与文澜阁本（后补抄），影响最大的是文渊阁本。

文渊阁／1925~1933年

Wenyuan Ge, 1925-1933

早在《四库全书》入藏文渊阁前，乾隆帝即为文渊阁的落成撰写了《文渊阁记》一文，立碑于阁旁。文中记述了《四库全书》的成书原委、传抄经过以及选址文华殿的原因。特别指出了《四库全书》的辑定，是乾隆帝“枕经葄史，镜已牖民”的要求，借以达到他“继绳祖考觉世之殷心，化育民物返古之深意”的目的。

文渊阁二层 / 1925~1933年

Second Floor of Wenyuan Ge, 1925-1933

文渊阁二层明间悬挂乾隆帝御笔“汇流澂鉴”金漆云龙匾额和楹联。文渊阁自乾隆四十一年（1776 年）建成后，皇帝每年在此举行经筵活动。四十七年（1782 年）《四库全书》告成之时，乾隆帝在文渊阁宴赏编纂《四库全书》的各级官员和参加人员，盛况空前。

文渊阁明间内景 / 1925~1933年

Interior Scene of the Bright Room in Wenyuan Ge, 1925-1933

《四库全书》连同《钦定古今图书集成》入藏文渊阁，按经、史、子、集四部分架放置。以经部儒家经典为首共22架和《四库全书总目考证》、《钦定古今图书集成》放置一层，并在中间设皇帝宝座，为讲经筵之处。二层中三间与一层相通，周围设楼板，置书架，放史部书33架。二层为暗层，光线极弱，只能藏书，不利阅览。三层除西尽间为楼梯间外，其它五间连通，每间依前后柱位列书架间隔，宽敞明亮。子部书22架、集部书28架存放在此。明间设御榻，备皇帝随时登阁览阅。乾隆皇帝为有如此豪华的藏书规模感到骄傲，曾作诗曰："丙申高阁秩干歌，今喜书成邺架罗。"清宫规定，大臣官员之中如有嗜好古书，勤于学习者，经允许可以到阁中阅览书籍，但不得损害书籍，更不许携带书籍出阁。

武英殿 / 1925~1949年

Wuying Dian (Hall of Military Prowess), 1925-1949

武英殿 / 1925~1949年

Wuying Dian, 1925-1949

明初帝王斋居、召见大臣皆在武英殿，崇祯时皇后千秋节命妇于此行朝贺礼。李自成入北京后，在武英殿当上了皇帝。清康熙年间武英殿开书局，置修书处，二百余年来刊印了大量印装精美的“殿本”图书。

断虹桥 / 1900年

Duanhong Bridge (Broken-Rainbow Bridge), 1900

断虹桥在武英殿东侧，南北向跨于金水河上。桥面铺砌汉白玉巨石，两侧石栏板雕穿花龙纹图案，望柱上之石狮神态各异，宛然如生。此桥用料之考究、装饰之华丽、雕刻之精美乃紫禁城诸桥之冠。断虹桥始建于元代（一说明初），是紫禁城中最古老的石桥。

西华门北侧城墙修缮 / 20世纪20年代

The Repairing Progress of North Side of Xihua Men, 1920s

照片中西华门东侧洋楼为古物陈列所时期修建的宝蕴楼。

南薰殿 / 20世纪50年代

Nanxun Dian (Hall of Southern Fragrance), 1950s

南薰殿位于武英殿西南，为一独立的院落，是明清两代贮藏帝后、贤臣画像的场所。殿内明、次间各设朱红漆木阁，分五层，安奉历代帝王像。殿之东室安奉历代皇后像，西室放置一木柜，贮明代帝后册宝。南薰殿明间立有乾隆帝《御制南薰殿奉藏图像记》卧碑，碑文中详细记载了殿内尊藏图像的情况：自太昊、伏羲以下，共有帝王贤臣画像（卷、册、轴）共121份，所绘大小人像共583名。

西华门 / 1900年

Xihua Men (West Glorious Gate), 1900

西华门是紫禁城四门之一，明清帝王至西苑和西郊园林游幸多从此门出。乾隆年间，为庆祝崇庆皇太后和乾隆帝万寿，特从紫禁城西华门开始，沿途搭设彩棚点景，直到西郊的皇家园林。

《崇庆皇太后万寿庆典图》卷中的西华门及门外点景 / 故宫博物院藏

Xihua Men and the Exterior Scene of Xihua Men from the Painting of Empress Dowager Chongqing's Birthday, The Palace Museum

东华门 / 1900年

Donghua Men (East Glorious Gate), 1900

东华门是紫禁城四门之一，清初东华门只准内阁官员出入，乾隆中期特许高年一、二品大员出入。遇大丧期间，皇帝、皇后、皇太后梓宫均从东华门出，遂其门钉为八路九列七十二颗，与其余三门不同。

角楼 / 约1907年

Corner Tower, Approx. 1907

角楼是紫禁城城池建筑的重要组成部分，它与紫禁城城墙、护城河、城门同属皇宫的防卫设施。角楼建筑形制精美别致，素有“九梁十八柱七十二脊”的说法。它的结构特征继承了我国古代木结构建筑灵活多变的传统做法，将使用功能和装饰效果完美地结合在一起。

角楼鸱吻 / 1956年

Chiwen (a dragon snout on both ends of a roof ridge) from Corner Tower, 1956

紫禁城西南角楼 / 20世纪初

The South-West Corner Tower of the Forbidden City, Early 20th century

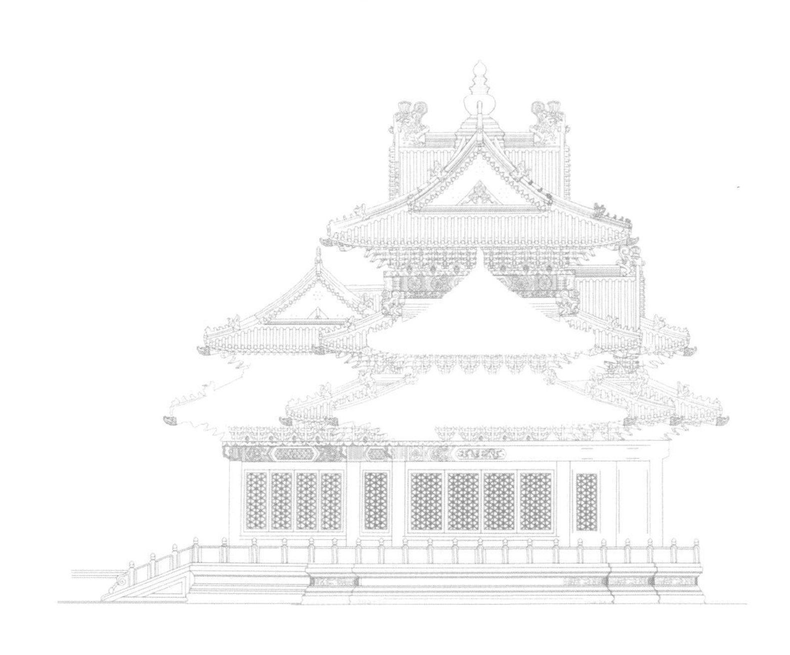

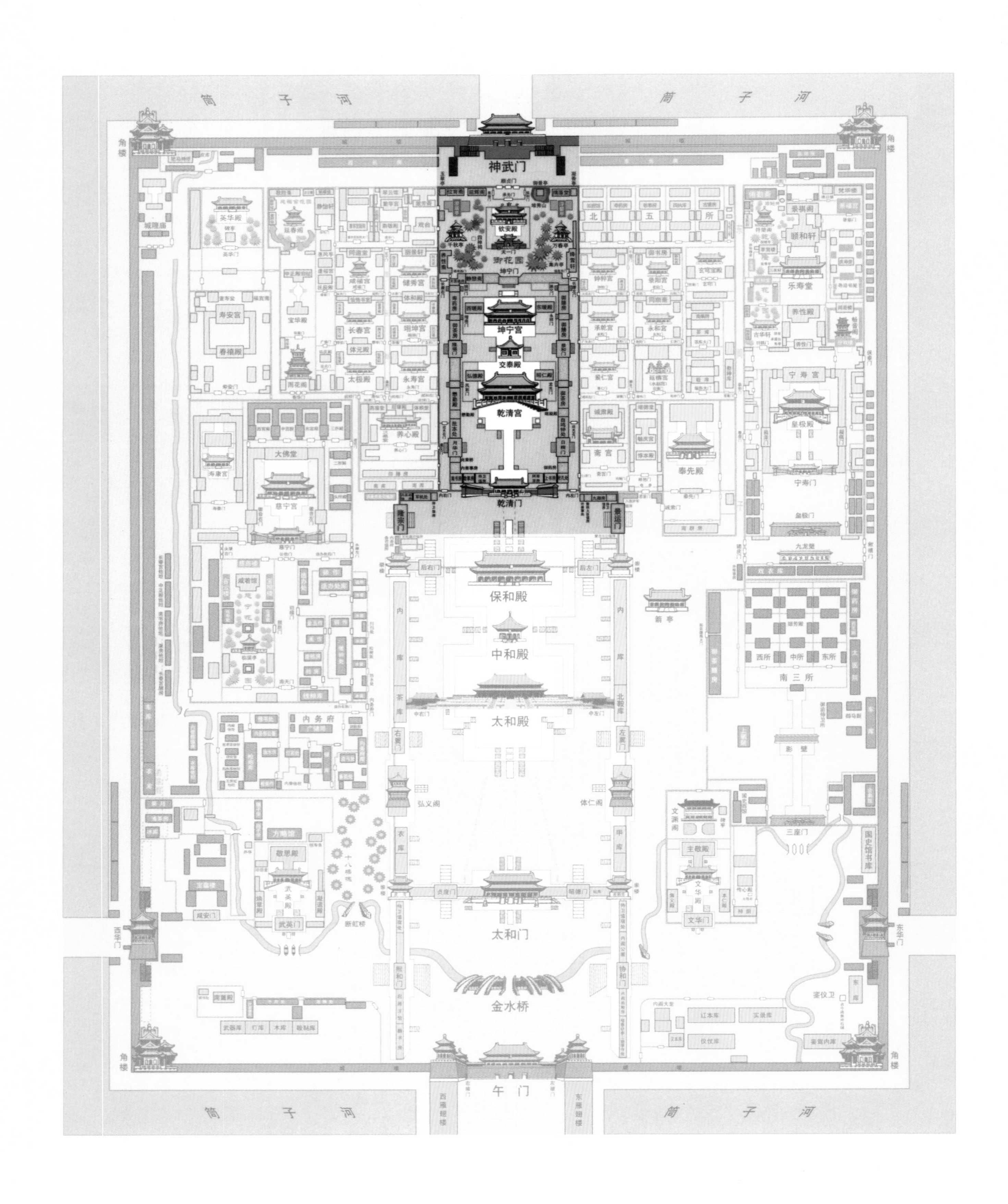

筒子河
角楼
神武门
御花园
钦安殿
坤宁宫
交泰殿
乾清宫
乾清门
隆宗门
景运门
保和殿
中和殿
太和殿
太和门
金水桥
午门
北五所
乐寿堂
宁寿宫
皇极殿
宁寿门
皇极门
九龙壁
南三所
西所
中所
东所
奉先殿
斋宫
养心殿
寿康宫
慈宁宫
大佛堂
寿安宫
英华殿
城隍庙
长春宫
永寿宫
承乾宫
永和宫
景仁宫
储秀宫
咸福宫
翊坤宫
体和殿
太极殿
延禧宫
钟粹宫
景阳宫
内务府
武英殿
武英门
断虹桥
文渊阁
文华殿
文华门
主敬殿
国史馆书库
东华门
西华门
弘义阁
体仁阁
右翼门
左翼门
协和门
熙和门
西雨檐楼
东雁翅楼
箭亭
影壁
三座门
颐和轩
养性殿

紫禁城内廷中路

The Middle Section of Inner Court of the Forbidden City

紫禁城内廷建筑规制比外朝要低，
注重等级的象征性与生活的实用性的结合。
内廷中路诸建筑以乾清宫为中心，由乾清门始，
依次是乾清宫、交泰殿、坤宁宫、坤宁门、御花园、
天一门、钦安殿、承光门、顺贞门、神武门。
皇帝在内廷中路的乾清宫处理政务与生活起居；
东西两庑有大臣的值所南书房和皇子学习的上书房；
皇后在交泰殿里接受千秋贺；
明代皇后的寝宫坤宁宫在清代被改成萨满祭祀处和帝后大婚洞房；
御花园是紫禁城四座花园中面积最大的一座，
分布着二十余座亭阁楼台，是宫廷游幸的御苑。

乾清门 / 20世纪初

Qianqing Men (Gate of Heavenly Court), Early 20th century

乾清门是紫禁城内廷正门，清代帝王举行“御门听政”的场所。听政时，设皇帝宝座于门内正中，“部院以次咨事，内阁面承谕旨”。乾清门西庑与东庑分别是翰林官员的值宿南书房和诸皇子读书的上书房。门前东西向“横街”，将紫禁城分为南部外朝、北部内廷两个区域。

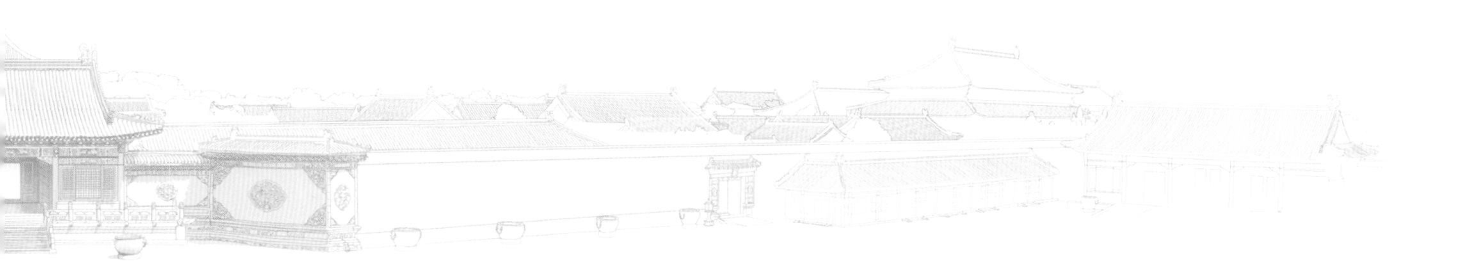

乾清门／1900年

Qianqing Men (Gate of Heavenly Court), 1900 century

乾清门前鎏金铜狮 / 1900年

A Golden Lacquer Bronze Lion in front of Qianqing Men, 1900

乾清宫 / 20世纪初

Qianqing Gong (Palace of Heavenly Court), Early 20th Century

乾清宫是明代皇帝在紫禁城中最重要的生活起居处。自雍正皇帝移居养心殿后，乾清宫作为皇帝召见大臣、批阅奏章、接见外藩使节和岁时受贺、筵宴的场所。皇帝“三大节”多御此殿受后妃宗室贺。

乾清宫 / 1922年
Qianqing Gong, 1922

乾清宫丹墀东侧 / 1922年

East Side of Dan Chi (platform) of Qianqing Gong, 1922

此帧照片所示的乾清宫丹墀上，跪拜着一群官员，殿前皇帝仪仗已经准备就绪，檐下还陈设着宫廷乐器。这是小朝廷时期一场“朝会”后，溥仪准备离开乾清宫的纪实照片。

乾清宫前铜镏金江山社稷金殿 / 1900年

The Golden Lacquer Bronze Hall of Jiangshan Sheji in front of Qianqing Gong, 1900

乾清宫前铜仙鹤／1900年

A Bronze Crane in front of Qianqing Gong, 1900

乾清宫前铜龟／1900年

A Bronze Turtle in front of Qianqing Gong, 1900

乾清宫前铜香炉 / 1900年
A Bronze Xiang Lu (Chinese thurible) in front of Qianqing Gong, 1900

乾清宫节日期间搭设的彩棚 / 约1922年

A Festive Staging in Qianqing Gong, Approx. 1922

乾清宫彩棚 / 约1922年

Festive Decorations in Qianqing Gong, Approx. 1922

乾清宫彩棚／约1922年

Festive Decorations in Qianqing Gong, Approx. 1922

乾清宫廊下安设的宫廷乐器 / 约1922年
The Imperial Instruments under the Colonnade of Qianqing Gong, Approx. 1922

乾清宫廊下安设的宫廷乐器 / 约1922年
The Imperial Instruments under the Colonnade of Qianqing Gong, Approx. 1922

乾清宫廊下安设的宫廷乐器／约1922年

The Imperial Instruments under the Colonnade of Qianqing Gong, Approx. 1922

乾清宫内景／1900年

Interior Scene of Qianqing Gong, 1900

乾清宫内景／20世纪初

Interior Scene of Qianqing Gong, Early 20th century

乾清宫明间设皇帝宝座，屏风、御案、香筒、仙鹤一应陈设俱全。殿内左右“列图史、玑衡、彝器”。昔日帝王御殿时，文武官员按官阶文东武西立于殿内外东西两侧。

乾清宫宝座上方悬“正大光明”匾额，为乾隆帝摹顺治帝御书。雍正帝继位后，清廷不再明宣皇太子人选，而是在紫禁城乾清宫“正大光明”匾后尊藏“建储锦匣”，待先皇崩逝后取出匣内御书姓名，明示储君，是为“秘密建储”制度。历史上，乾隆、嘉庆、道光、咸丰四位皇帝是以此方式继承帝位的。“秘密建储”制度有效地避免了皇子之间争夺储位的斗争。

乾清宫内景 / 20世纪初

Interior Scene of Qianqing Gong, Early 20th century

照片中乾清宫内悬挂着华丽的西洋玻璃灯。据清末太监耿进喜回忆，光绪二十六年（1900 年）始，紫禁城内开始安装电灯，发电装置设在北池子电灯局，紫禁城近五百年里第一次被电灯点亮。

乾清宫宝座 / 1900年

The Throne in Qianqing Gong, 1900

乾清宫宝座后的屏风上镌刻着康熙皇帝御制《五屏风铭》，中堂曰：惟天聪明，惟圣时宪，惟臣钦若，惟民从乂；东一堂曰：首出庶物，万国咸宁；西一堂曰：恺悌君子，四方为则；东次堂曰：功崇惟志，业广为勤；西次堂曰：知人则哲，安民则惠。

乾清宫雕云龙纹镜 / 1900年

A Grand Sandalwood Mirror in Qianqing Gong, 1900

交泰殿 / 20世纪初

Jiaotai Dian (Hall of Union), Early 20th century

清代每逢元旦、千秋二节，皇后于此升座，接受宫中妃嫔、命妇的朝贺。乾隆皇帝钦定“二十五宝”后，将宝玺存放在殿内宝座周围。每年正月，由钦天监选择吉日吉时，设案开封陈宝，皇帝来此拈香行礼。顺治皇帝所立“内宫不许干预政事”的铁牌曾立于此殿。皇帝大婚时，皇后的册、宝安设殿内左右案上。每年春季祀先蚕，皇后先一日在此阅视采桑的用具。照片中交泰殿东侧阶下，沿后三宫台基建有一排低矮的小屋，这里曾是旧时值事人员的“他坦”房，生活气息十分浓厚。此外，照片中乾清宫东西三间后檐下曾砌有矮墙一道，当是为约束外间进出内廷中路所设，与今日格局不同。

乾清宫东暖阁 / 20世纪初

The East Warm Chamber of Qianqing Gong, Early 20th century

照片中乾清宫东暖阁御座上方悬挂着乾隆帝御制《乾清宫铭》和道光帝《辛卯元旦自述》。宝座后的“仙楼”格局清晰可见。

节日时的交泰殿 / 约1922年

Festive Decorations of Jiaotai Dian, Approx. 1922

节日时交泰殿张贴门神／约1922年

The Poster of Door God Placed on Jiaotai Dian, Approx. 1922

交泰殿内蟠龙藻井 / 1900年

The Caisson of Pan in Jiaotai Dian, 1900

交泰殿内景／1900年

Interior Scene of Jiaotai Dian, 1900

交泰殿内“无为”匾额曾是康熙帝御笔，但额上落款处“乾隆六十二年丁巳御笔恭摹”字样赫然在目。原来，乾清宫、交泰殿在乾隆帝禅位的第二年被火焚毁，并于当年重建。原康熙帝御笔匾额无存，现存是乾隆帝摹写后悬挂上去的，至于其摹所据何本，就不得而知了。“乾隆六十二年”的字样仅限于内廷，实际上这一年是嘉庆二年（1797年）。“无为”额下有乾隆帝御书《交泰殿铭》，“二十五宝”就存放在宝座周围的宝函内。

交泰殿内铜壶滴漏 / 约1922年

A Bronze Water Clock in Jiaotai Dian, Approx. 1922

交泰殿内大自鸣钟 / 20世纪初

A Giant Striking Clock in Jiaotai Dian, Early 20th Century

交泰殿内陈设有计时器二座，分别是殿内东侧的铜壶滴漏和西侧的大自鸣钟，均制作于嘉庆年间。此大自鸣钟为宫中计时的标准，有太监专门负责上弦、洒扫。旧时，此钟报时后神武门鸣钟鼓，城外钟鼓楼亦随之响应。

关于自鸣钟，明末宫廷始见，最初为觐谒皇帝的传教士从西洋携来。到清代，内府通过粤海关向英国订购自鸣钟的同时，在北京的清宫造办处还设有专门的“做钟处”，专司成造、修理钟表。此外，苏州、广州一带的钟表制作也达到了相当的水平。在清代，自鸣钟是宫苑中的重要陈设品，甚至将其安设在建筑之上。仅在圆明园焚毁后，内府向皇帝呈报的失窃钟表就有400余座之多，可见清帝对自鸣钟的喜爱。

圆明园慈云普护钟楼／1744年写景图

Clock Tower of Old Summer Palace from Ciyun Puhu,
A Painting of Landscape in 1744

乾隆年间，圆明园慈云普护曾建有一座六角钟楼，二层安设一架金漆大表盘，名“时时如意时刻钟”，用于园内报时。此类钟楼在南海瀛台（见本书257页）、清漪园（见本书284页）、静宜园中都有兴建。

坤宁宫 / 20世纪初

Kunning Gong (Palace of Earthly Tranquility), Early 20th century

坤宁宫在明代是皇后寝宫，清代改为萨满祭祀场所。旧时，坤宁宫日日行祭祀礼，分为朝祭和夕祭；另有春秋大祭、求福祭等。每岁腊月二十三，坤宁宫祀灶，届时要奏请皇帝于佛前、神前、灶前拈香行礼。

照片上坤宁宫前竖立着“索伦杆”，萨满祭神时，将大锅蒸煮后的“神肉”置于锡斗中，由“索伦杆”高举上天，供“神鸟”享用。

坤宁宫内萨满祭神处／约1922年

The Room for Shanmanist Worship of Kunning Gong, Approx. 1922

清代改建祭祀场所以后，将坤宁宫明间宫门封死，改在东次间开门，西侧室内砌满族传统的炕床与煮肉大锅，又设供案于殿内。祭祀时，大锅煮起神肉，盘盛整块方肉分给众人。据曾在坤宁宫吃过肉的大臣回忆，神肉没有佐料，腥膻无比，有人事先将盐粒藏在衣袖里，待分到肉后偷偷蘸盐才可勉强下咽。

坤宁宫节日时搭设的彩棚／约1922年

A Festive Staging in Kunning Gong, Approx. 1922

坤宁宫东三间曾是康熙、同治、光绪帝大婚时的洞房，1922 年逊帝溥仪大婚也将洞房设在这里。

坤宁宫东暖阁喜床／约1922年
The Wedding Bed of East Warm Chamber in Kunning Gong, Approx. 1922

坤宁宫东暖殿明间内景／约1922年
Interior Scene of East Warm Chamber in Kunning Gong, Approx. 1922

“喜”字屏风上悬挂雍正帝御书“位正坤元”金漆云龙匾额，意示皇后位居坤宫正位。

坤宁宫东暖阁东间内景／约1922年
Interior Scene of East Room in East Warm Chamber of Kunning Gong, Approx. 1922

坤宁宫东暖殿东间内景／约1922年
Interior Scene of East Room in East Warm Chamber of Kunning Gong, Approx. 1922

坤宁宫东暖殿
东间陈设／约1922年
The Interior Design of East Room in East Warm Chamber of Kunning Gong, Approx. 1922

坤宁宫东暖殿
西间陈设／约1922年
The Interior Design of West Room in East Warm Chamber of Kunning Gong, Approx. 1922

坤宁宫东暖阁毗卢帽上“日升月恒”匾额／约1922年

A Tablet with Risheng Yueheng in East Warm Hall of Kunning Gong, Approx. 1922

坤宁宫东暖阁双开木板“喜”字门／约1922年

The Door with the Character Xi in East Warm Hall of Kunning Gong, Approx. 1922

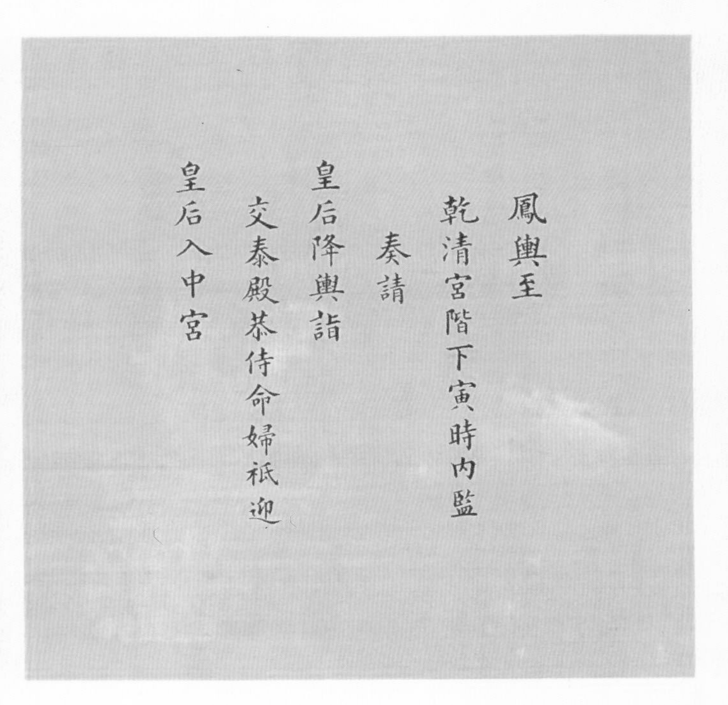

鳳輿至
乾清宮階下寅時內監
奏請
皇后降輿詣
交泰殿恭侍命婦祗迎
皇后入中宮

庆宽绘《载湉大婚典礼全图》中的交泰殿与坤宁宫
故宫博物院藏

Kunning Gong and Jiaotai Dian in Paining of the Imperial Wedding of Zaitian (Guangxu Emperor), Drew by Qingkuan, The Palace Museum

天一门 / 1922年

Tianyi Men (Tianyi Gate), 1922

天一门砖石结构，灰墙黄瓦，是钦安殿院落正门。名取“天一生水，地六承之”，以应钦安殿主尊玄武大帝北方水神之意。

天一门内连理柏 / 1925~1949年

A Cypress Tree Looks Like a Couple Holding Hands, 1925-1949

天一门前“海参”石 / 1925~1949年

A Stone of Sea Cucumbers in front of Tianyi Men, 1925-1949

天一门前“诸葛拜斗”石 / 1925~1949年

A Stone of Zhuge Baidou in front of Tianyi Men, 1925-1949

钦安殿 / 1922年

Qin'an Dian (Hall of Imperial Peace), 1922

钦安殿是紫禁城中轴线上的一座宗教建筑，内供玄武大帝。旧时，每岁元旦，宫殿监率该处首领太监等于天一门内正中设斗香，恭候皇帝亲诣拈香行礼；每月朔、望日，由宫殿监等敬谨拈香行礼。

钦安殿抱厦内梁枋彩绘 / 1900年

The Painting on the Timber Beam of Qin'an Dian, 1900

钦安殿后檐墙与承光门 / 1925~1949年

The Back Eaves of Qin'an Dian and Chengguang Men, 1925-1949

绛雪轩 / 1917~1927年

Jiangxue Xuan (Pavilion of Crimson Snow Flakes), 1917-1927

绛雪轩形作“凸”字，在御花园与养性斋东西相对。旧时，因轩前植有五株海棠，每到落花时节，海棠花纷飞若雪，遂得“绛雪”之名。与今天不同的是，照片中的绛雪轩被一组巨大的凉棚遮盖，在20世纪40年代的测绘图上，标示出这架凉棚的高度超过了10米，风格中西合壁。

绛雪轩立面图 / 20世纪40年代测绘

An Architectural Drawing of Jiangxue Xuan, Measured and Drew in 1940s

绛雪轩南山墙 / 1900年

South Gable of Jiangxue Xuan, 1900

绛雪轩前琉璃花台 / 1925~1949年

Glazed Platform in front of Jiangxue Xuan, 1925-1949

琉璃花台上植太平花，台前立露陈若干，其中有乾隆三十一年（1766年）黑龙江将军福僧阿进贡木变石（木化石）一座，上镌乾隆帝御制《咏木变石》诗一首：

不记投河日，宛逢变石年。
磕敲自铿尔，节理尚依然。
旁侧枝都谢，直长本自坚。
康干虽岁贡，逊此一峰全。

养性斋／1900年

Yangxing Zhai (Study of Nature Cultivation), 1900

养性斋形作“凹”字，在御花园与绛雪轩东西相对，曾是逊帝溥仪英文教师庄士敦住所。

养性斋内庄士敦书房 / 20世20年代

Study of Sir Reginald Fleming Johnston in Yangxing Zhai, 1920s

养性斋与四神祠 / 1925~1949年

Yangxing Zhai and Sishen Ci (Temple of Four Gods), 1925-1949

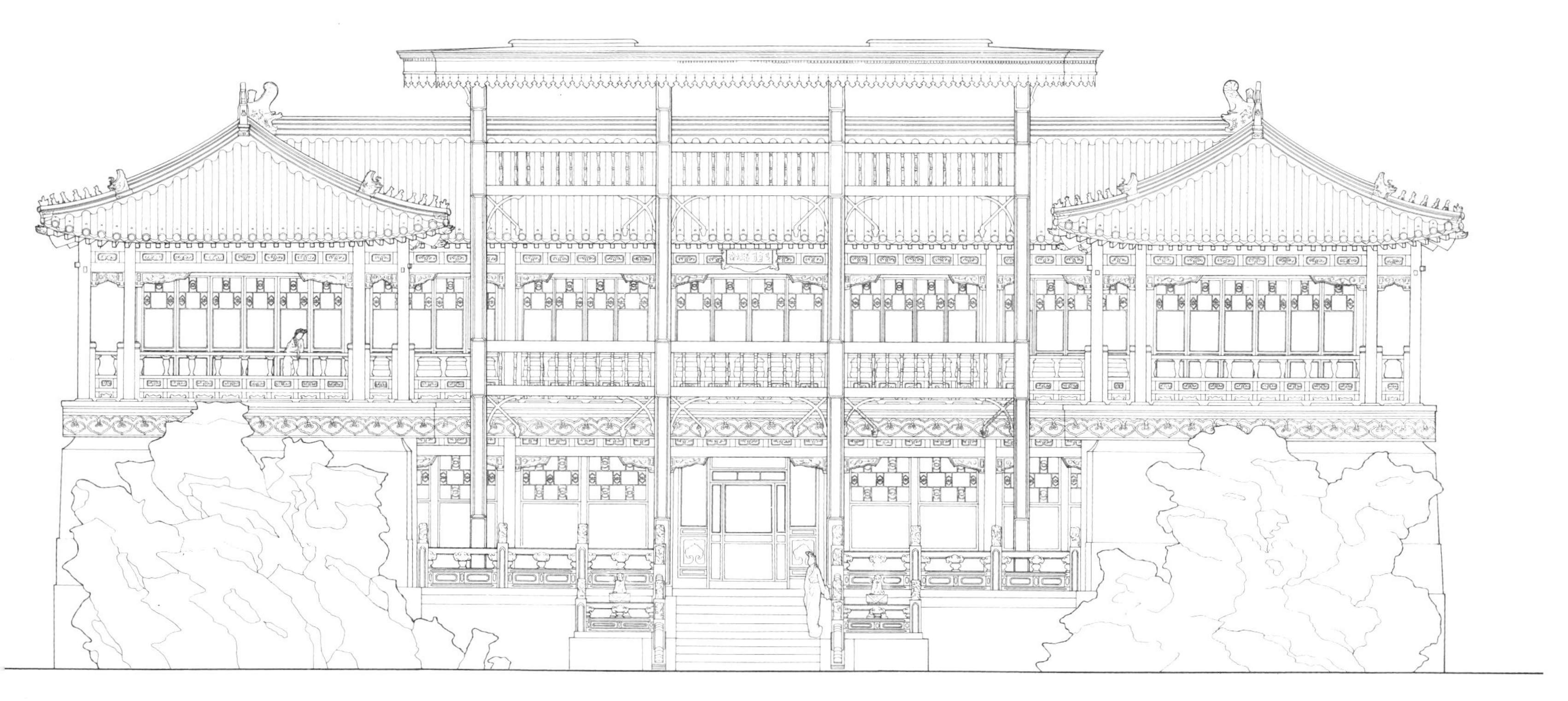

养性斋正立面图 / 20世纪40年代测绘

An Architectural Drawing of Yangxing Zhai, Measured and Drew in 1940s

养性斋前山石 / 20世纪初
Stone Hill of Yangxing Zhai, Early 20th Century

养性斋前山石 / 20世纪初
Stone Hill of Yangxing Zhai, Early 20th Century

养性斋前山石 / 20世纪初

Stone Hill of Yangxing Zhai, Early 20th Century

千秋亭 / 1917~1924年

Qianqiu Ting (Thousand-Autumn Pavilion), 1917-1924

千秋亭在御花园与万春亭东西相对，天圆地方形制，中设蟠龙藻井。亭内原供佛像，同治帝驾崩后曾供奉神牌于此。照片中千秋亭北侧的澄瑞亭与抱厦皆为封闭状态，与今天的敞轩形制不同。亭北位育斋。仔细观察照片，逊帝溥仪就站在千秋亭栏杆一旁。

千秋亭 / 1917~1924年

Qianqiu Ting, 1917-1924

万春亭 / 20世纪20年代

Wanchun Ting (Ten-Thousand-Spring Pavilion), 1920s

万春亭在御花园与千秋亭东西相对，天圆地方形制，中设蟠龙藻井。亭内曾供关公像。照片中的竹篱墙今已不存。

万春亭西侧 / 20世纪20年代

West Side of Wanchun Ting, 1920s

此帧照片中万春亭西侧的汉白玉台基上曾建有歇山顶小殿一座，名“禊赏”，今已不存。淑妃文绣就藏在禊赏亭旁的柏树下。

浮碧亭与摛藻堂／20世纪20年代

Fubi Ting (Pavilion of Green Ripples) and Chizao Tang (Hall of Literary Elegance), 1920s

浮碧亭在御花园与澄瑞亭东西相对，照片中浮碧亭内檐曾满悬匾额。亭北摛藻堂，作为乾隆皇帝的书房曾收贮《四库全书荟要》一万二千册，依《四库全书》之序，按经、史、子、集分藏于堂内东西书架上。摛藻堂外，西墙壁上镌刻有乾隆帝御制《古柏行》诗一首。

澄瑞亭东侧／20世纪20年代

East side of Chengrui Ting (Pavilion Auspicious Clarity), 1920s

澄瑞亭在御花园与浮碧亭东西相对，照片中澄瑞亭内檐曾满悬匾额。尚可辨认出“鱼跃鸢飞”、“与物皆忘”。照片中间和左侧站立者是皇后婉容和淑妃文绣，右立者是溥杰妻唐石霞。

御花园鹿囿／20世纪20年代

The Deer Garden of Yuhua Yuan, 1920s

御花园中曾设鹿囿一座，位于养性斋东，与绛雪轩西侧的集卉亭（今已不存）相对。旧时宫中圈养鹤鹿，取“鹤鹿同春”之意。鹿囿北侧建有高台，台下开券门直通南侧的鹿圈。

御花园古柏 / 20世纪20年代

An Ancient Cypress Tree in Yuhua Yuan, 1920s

御花园西井亭 / 1922年
Xijing Ting of Yuhua Yuan, 1922

延晖阁／20世纪20年代

Yanhui Ge (Pavilion of Prolonged Sunshine), 1920s

延晖阁明代称清望阁，清代改今名，在御花园中与堆秀山东西相对。嘉庆时，嘉庆帝将乾隆帝遗留的御书存贮在这里。晚清时，延晖阁前曾作为内廷遴选八旗秀女的场所。

延晖阁立面图／20世纪40年代测绘

An Architectural Drawing of Yanhui Ge, Measured and Drew in 1940s

堆秀山／20世纪20年代

Duixiu Shan (Rockery Hill), 1920s

堆秀山万历时成山，原为明代观花殿旧址。山上筑亭名“御景”，是帝后重阳登高赏景之处。堆秀山中藏储水缸，与山下两座水法连通，可上演双龙吐水景观。

承光门 / 1900年

Chengguang Men (Gate of Inheriting Light), 1900

由堆秀山西望延晖阁 / 20世纪20年代

Yanhui Ge, from the angle of the west side of Duixiu Shan, 1920s

神武门 / 1900年

Shenwu Men (Gate of Divine Might), 1900

神武门为紫禁城北门，是紫禁城中保存至今的明代木结构建筑。照片所示为八国联军占领北京期间，日本侵略军与清兵在神武门前合影，此时的紫禁城北门由日本兵把守。

神武门 / 1928~1937年

Shenwu Men, 1928-1937

照片中神武门门洞上“故宫博物院”五字为李煜瀛 1925 年手书，原为木匾，后改为石匾。门前交叉的“青天白日旗”与“青天白日满地红旗”为我们明确了照片拍摄的时间上限，即 1928 年北伐结束，国民政府形式上统一全国以后，时间下限当在 1937 年 7 月北平沦陷以前。

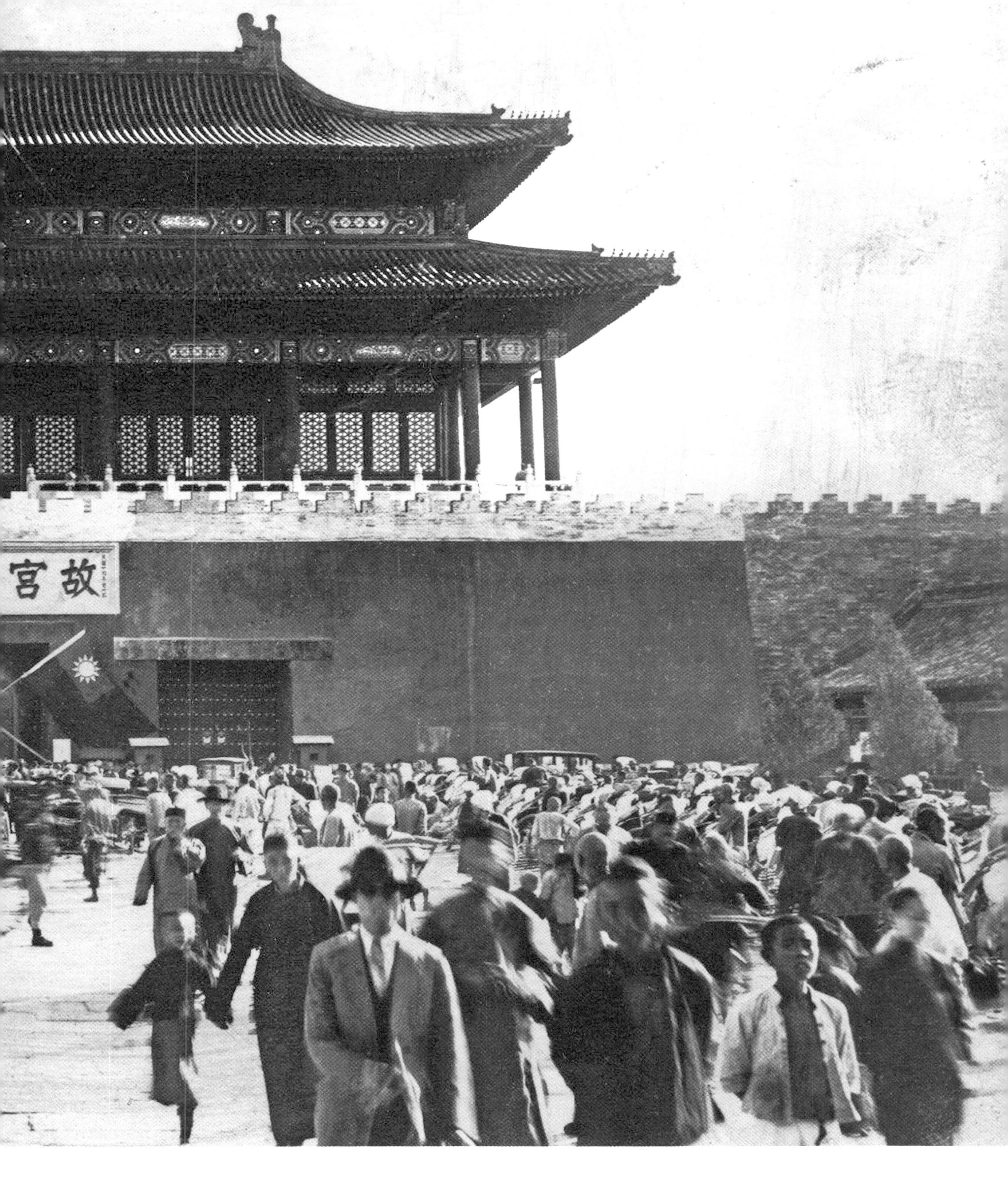
故宫

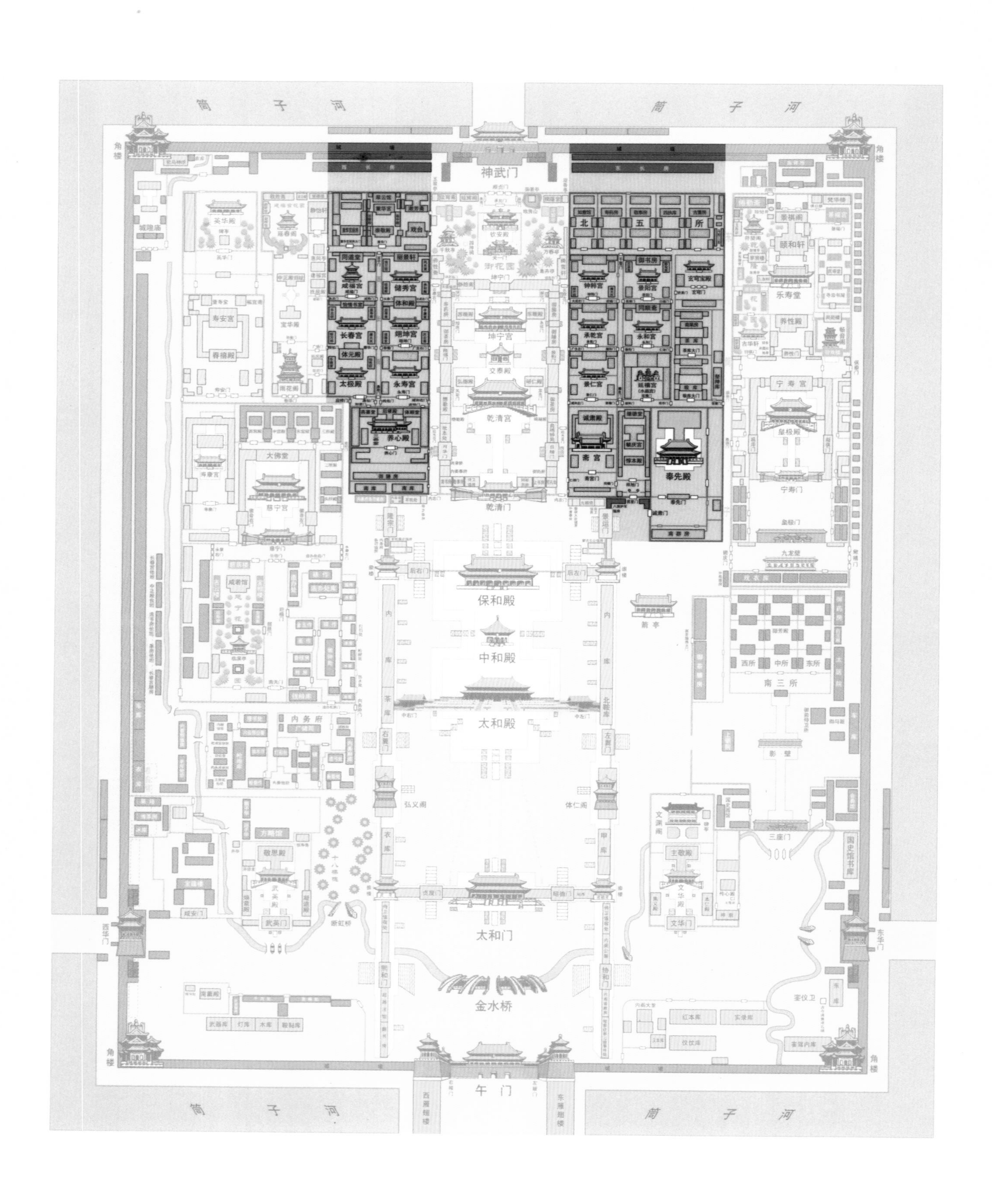

筒子河
筒子河
角楼
角楼
神武门
御花园
北五所
咸福宫
储秀宫
长春宫
翊坤宫
体元殿
太极殿
永寿宫
钟粹宫
景阳宫
承乾宫
永和宫
景仁宫
养心殿
斋宫
诚肃殿
奉先殿
坤宁宫
交泰殿
乾清宫
乾清门
英华殿
寿安宫
雨花阁
宝华殿
城隍庙
大佛堂
寿康宫
慈宁宫
颐和轩
乐寿堂
养性殿
宁寿宫
皇极殿
宁寿门
皇极门
九龙壁
隆宗门
景运门
后右门
后左门
保和殿
中和殿
太和殿
箭亭
西所
中所
东所
南三所
影壁
内务府
右翼门
左翼门
弘义阁
体仁阁
文渊阁
主敬殿
文华殿
文华门
三座门
国史馆书库
敬思殿
武英殿
武英门
断虹桥
西华门
东华门
太和门
熙和门
协和门
金水桥
銮仪卫
角楼
角楼
午门
筒子河
筒子河

紫禁城内廷东路、西路

The Eastern and Western Section of Inner Court of the Forbidden City

以中轴线为界，

紫禁城内廷东路和西路的主要建筑有东六宫、西六宫、

养心殿、斋宫、毓庆宫、奉先殿，

还包括了由原乾东五所和乾西五所演变而来的北五所和重华宫一区。

东、西六宫是皇帝赐居后妃的宫殿；

养心殿是皇帝在紫禁城内最重要的办公和就寝场所；

皇帝郊坛祭祀前在斋宫斋戒；

皇子的生活学习处在毓庆宫；

奉先殿是奉祀先朝帝后的“家庙”；

重华宫曾是乾隆帝的潜邸；

北五所内有如意馆、寿药房、四执库等内务府服务宫廷的机构和库房，

管理太监事务的敬事房也在这一区域。

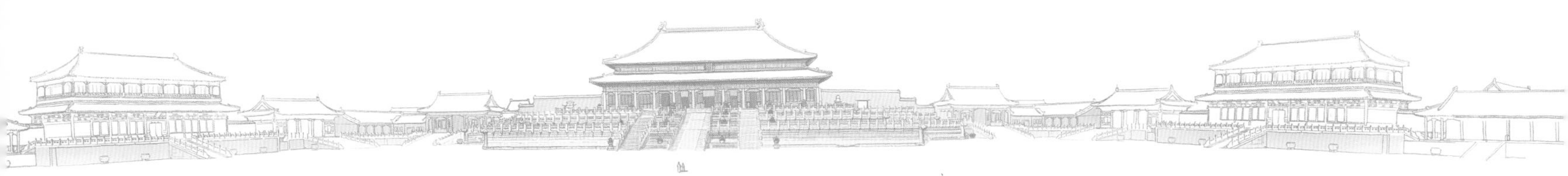

养心门 / 1922年

Yangxin Men (Gate of Mental Cultivation), 1922

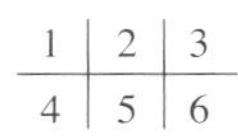

1.养心门外玉壁 / 1922年
Bronze Screen Wall with Jade outside the Yangxin Men, 1922

2.养心门内 / 1922年
Inside the Yangxin Men, 1922

3.养心殿抱厦 / 1922年
The Covered Corridor of Yangxin Dian, 1922

4.养心门外西值房 / 1922年
The West Guardian Room outside the Yangxin Men, 1922

5.养心殿抱厦 / 1922年
The Covered Corridor of Yangxin Dian, 1922

6.养心殿明间外景 / 1922年
Exterior Scene of the Bright Room in Yangxin Dian, 1922

自雍正帝始，清代历朝皇帝均以养心殿为处理政务、引见官员和生活起居的场所。养心殿与内右门外的军机处值房毗邻，外邻军机处，内通后三宫，且前殿与后殿呈“工”字格局，前殿办公，后殿就寝，这里又有独立的御膳房，这些条件都极大地方便了皇帝的公务和生活。

养心殿明间内景 / 1900年

Interior Scene of the Bring Room in Yangxin Dian, 1900

养心殿明间设御座，御座周围陈设着香筒、几案、翎扇、屏风，屏风两侧置书厨。明间上设蟠龙藻井，下铺苏松“金砖”，这里是清帝处理日常政务的重要场所。

照片中宝座上方的“中正仁和”匾额为雍正帝御书，周围匾额上书有清代历朝皇帝的圣训。宝座后的屏风上，雕刻着乾隆帝作于乾隆二十五年（1760 年）的《新正养心殿》诗。这一年正月，适逢准噶尔回部叛乱平定、清廷一举“拓地二万里”、将军兆惠班师凯旋之际，乾隆帝在养心殿作此诗以纪念自己的西师武功，借养心殿之名抒发其闲中养心、安不忘危、自强不息的心志。

新正養心殿

西師歸振旅東陲舒由庚執徐奉執規持躬凛持盈三白兆宿歲萬幾簡新正閒中足養心而予有所怦懷安豈良圖無逸斯元亨自強勵不息善長時偕行

乾隆帝御制《新正养心殿》诗

Poem on Xinzheng Yangxin Dian, Wrote by Qianlong Emperor

养心殿明间蟠龙藻井 / 1900年

The Caisson of Pan in Yangxin Dian, 1900

养心殿东暖阁内景／1900年

Interior Scene of the East Warm Chamber in Yangxin Dian, 1900

养心殿东暖阁是清帝朝后引见官员、批阅奏章之处。1861年“辛酉政变”后，清廷在东暖阁东侧西向设御座，引见时同治帝坐于前（西），慈安、慈禧两宫太后坐于后（东），中间悬挂帐幕，此即著名的“垂帘听政”处。

养心殿东暖阁内景／20世纪20年代
Interior Scene of the East Warm Chamber in Yangxin Dian, 1920s

养心殿东暖阁内景／20世纪20年代
Interior Scene of the East Warm Chamber in Yangxin Dian, 1920s

养心殿随安室/ 1900年

Suian Shi of Yangxin Dian, 1900

此帧照片中的随安室还保留着慈禧太后与光绪帝西逃前的原状。

养心殿东暖阁内景
20世纪20年代
Interior Scene of the East Warm Chamber in Yangxin Dian, 1920s

养心殿体顺堂内景
20世纪20年代
Interior Scene of Tishun Tang in Yangxin Dian, 1920s

体顺堂在养心殿后殿东，西与燕禧堂相对，是清代皇后侍寝时的暂住寝宫。照片中“一堂喜气”匾为慈禧太后御书。

养心殿寝宫内景 / 20世纪20年代
Interior Scene of the Bedroom in Yangxin Dian, 1920s

养心殿东暖阁前檐炕床 / 20世纪20年代
A Kang (a brick bed) at the East Warm Chamber of Yangxin Dian, 1920s

养心殿院内花卉 / 20世纪20年代
The Flowers at the Yard of Yangxin Dian, 1920s

养心殿院内花卉 / 20世纪20年代
The Flowers at the Yard of Yangxin Dian, 1920s

养心殿院内花卉 / 20世纪20年代
The Flowers at the Yard of Yangxin Dian, 1920s

斋宫抱厦内景 / 1900年
Interior Scene under the Covered Corridor of Zhai Gong, 1900

斋宫明间内景 / 1900年
Interior Scene of the Bright Room in Zhai Gong, 1900

斋宫殿内陈设 / 1900年
The Interior Design of Zhai Gong, 1900

斋宫 / 20世纪初

Zhai Gong (Palace of Abstinence), Early 20th Century

斋宫建于雍正九年（1731 年），是清帝在天地、祈谷、常雩大祀礼前，在紫禁城内斋戒的地方。旧时，遇皇帝宿斋宫，宫殿监恭设斋戒牌、铜人于斋宫丹陛左侧。斋戒日，皇帝与陪祀大臣佩戴斋戒牌，各宫悬斋戒牌于帘额之上。斋戒期间，宫中不作乐，不饮酒，忌食辛辣。

毓庆宫东里间内景 / 20世纪20年代

Interior Scene of the East Room in Yuqing Gong, 1920s

毓庆宫康熙时曾归允礽居住，其后一直作为皇子寝宫使用。由于乾隆帝禅位嘉庆帝后，并没有搬出养心殿，所以嘉庆帝以皇帝身份在此宫居住了三年时间。这里也是逊帝溥仪幼时读书的地方。

承乾宫 / 1925~1949年

Chengqian Gong (Palace of Inheriting Heaven), 1925-1949

承乾宫为东六宫之一，明代称永宁宫，崇祯时改今名。曾是顺治帝宠妃董鄂氏的寝宫；咸丰帝生母孝全皇后曾居住于此。

永和宫抱厦 / 20世纪20年代

The Covered Corridor of Yonghe Gong (Palace of Eternal Harmony), 1920s

永和宫抱厦下的鹦鹉 / 20世纪20年代

A Cockatoo under the Covered Corridor of Yonghe Gong, 1920s

永和宫为东六宫之一，明初称永安宫。曾是雍正帝生母德妃乌雅氏的寝宫；道光帝的静妃在此生下了恭亲王奕䜣；光绪帝的瑾妃亦在此宫居住过。

永和宫后殿同顺斋内景 / 1925~1949年

Interior Scene of Tongshun Zhai, 1925-1949

在此帧照片中，同顺斋明间悬有御书“慎修思永”、“福寿之徵”匾。墙壁上尚悬挂着婉容、瑾妃等逊清人物照片。同顺斋明间摆放着大量自鸣钟，均系着带有“国立故宫博物院”字样的标签，可知照片的拍摄时间应在1925年故宫博物院成立后、对清宫旧藏文物点查后的某一时间。

同顺斋婉容、文绣与溥仪姊弟们的合影 / 1922年

A Photograph of Wanrong, Wenxiu and Siblings of Puyi, 1922

翊坤宫内景 / 1900年

Interior Scene of Yikun Gong
(Palace of Blessings to Mother Earth), 1900

翊坤宫前青铜露陈 / 1900年

Bronze Lu Chen in front of Yikun Gong, 1900

翊坤宫为西六宫之一，明代称万安宫。万历宠妃郑氏曾居住于此。照片中翊坤宫廊下陈设着自鸣钟。

体和殿东配殿寝床 / 20世纪初

A Bed in the East Wing Hall of Tihe Dian (Hall of State Harmony), Early 20th Century

慈禧太后居储秀宫时，通常会在体和殿进膳。在这里慈禧太后为光绪帝“挑选了”一后二妃。

储秀宫 / 1900年

Chuxiu Gong (Palace of Gathering Excellence), 1900

储秀宫为西六宫之一，明初称寿昌宫。光绪十年（1884 年）慈禧太后花费 63 万两白银将储秀宫装修一新，在此度过了她的五十岁寿辰。

储秀宫青铜露陈
1900年
Bronze Lu Chen of Chuxiu Gong, 1900

储秀宫南窗炕床
20世纪20年代
A Kang at the South Window of Chuxiu Gong, 1920s

储秀宫内浴缸
20世纪20年代
A Bathtub in Chuxiu Gong, 1920s

储秀宫 / 1924~1927年

Chuxiu Gong, 1924-1927

体元殿后抱厦（长春宫院内戏台） / 1900年

The Covered Corridor behind the Tiyuan Dian (Hall of State Unity), 1900

长春宫内景 / 20世纪20年代

Interior Scene of Changchun Gong (Palace of Long-Spring-Time), 1920s

长春宫为西六宫之一，嘉靖时一度称永宁宫。这里曾是乾隆帝孝贤皇后的寝宫。皇后去世后，乾隆帝命内廷保持长春宫皇后生活时的原状。清末慈禧太后曾居此，逊帝溥仪的淑妃文绣亦生活在长春宫。

西二长街嘉祉门 / 20世纪20年代

Jiazhi Men at the Xi'er Long Street, 1920s

嘉祉门上安装了电灯，可知此帧照片拍摄于光绪二十六年（1900 年）以后。

崇敬殿东间 / 20世纪20年代

East Room of Chongjing Dian (Hall of Adoration), 1920s

崇敬殿是重华宫前院正殿，明间悬挂乾隆帝为宝亲王时书写的“乐善堂”黑漆金字匾，这里是弘历大婚后登基前的住所。殿后重华宫，每年新正清帝在此赐内廷大学士、翰林茶宴，席间君臣赋诗联句，引为家法。

重华宫东间陈设 / 20世纪20年代

The Interior Design of the East Room in Chonghua Gong (Palace of Double Brilliance), 1920s

殿内寿字立轴上悬挂着慈禧太后御书“笃礼崇义”匾，北墙上挂一黑漆金字挂屏，上刻御书唐李乂《陪幸临渭亭遇雪应制》诗。殿内多宝格炕几后面，张挂着雪景行乐题材的通景屏。

重华宫东间陈设 / 20世纪20年代

The Interior Design of the East Room in Chonghua Gong (Palace of Double Brilliance), 1920s

漱芳斋外景/ 20世纪20年代

Exterior Scene of Shufang Zhai, 1920s

照片所示漱芳斋外檐装饰着复杂的西洋蕃草、中式砖雕和西洋铁艺，廊柱上描绘着牡丹花，戏台前还安设了西洋路灯，颇显近代味道。

漱芳斋是清帝宴集观戏之所。每年元旦，清帝要在此举行开笔仪式。元旦这天，皇帝着吉服于殿内升座，御案上陈设着金瓯永固杯，杯中注满屠苏酒，皇帝拿起“赐福苍生”笔，书就新年的第一字，是为“开笔”。开笔所书往往是福、禄、寿、喜一类的吉祥文字，将其赏赐大臣、宗室、外藩，领受人视为无上的荣耀。

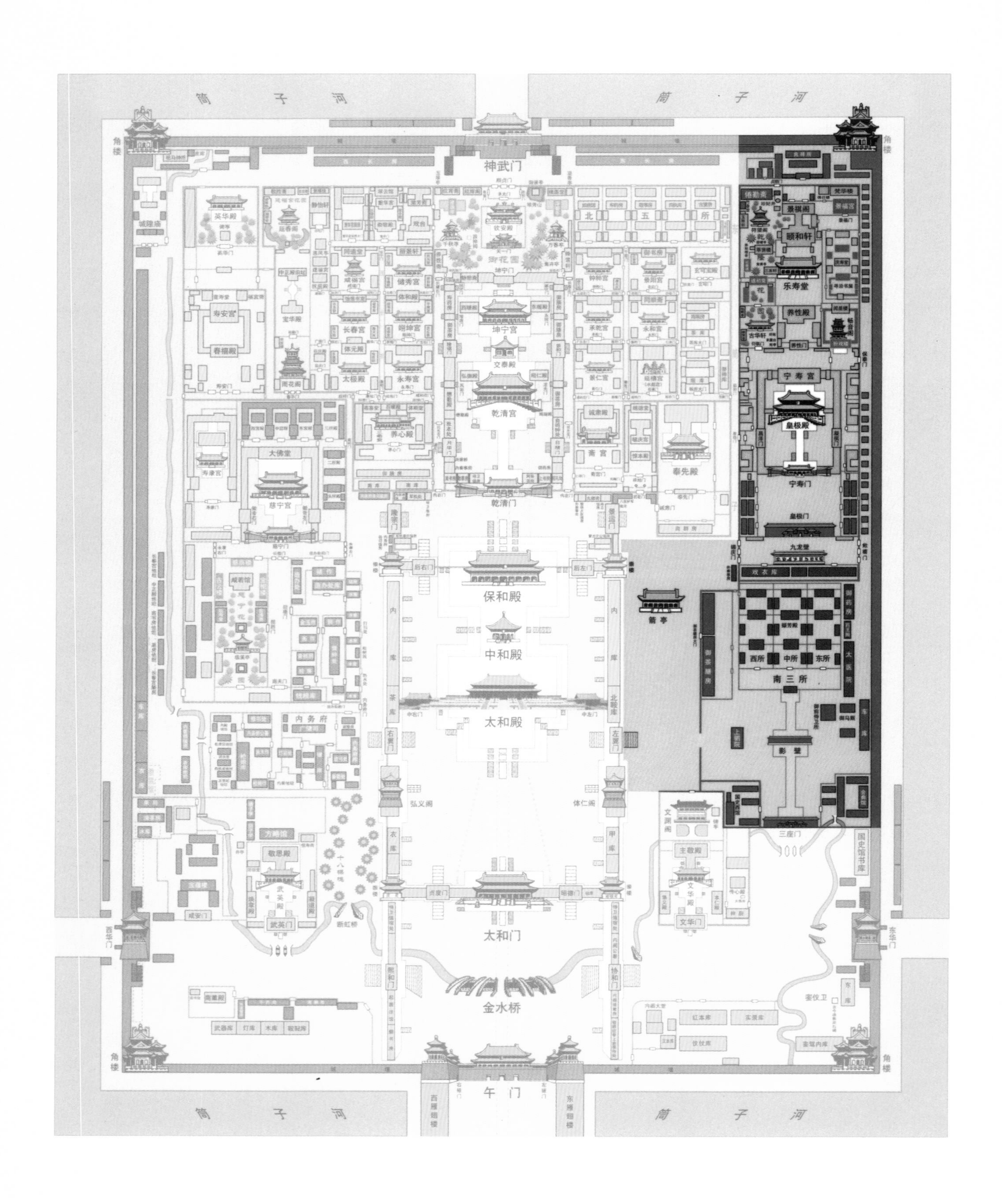

筒子河
筒子河
角楼
角楼
神武门
北五所
英华殿
钦安殿
御花园
坤宁门
储秀宫
翊坤宫
长春宫
太极殿
永寿宫
坤宁宫
交泰殿
乾清宫
寿安宫
春禧殿
雨花阁
养心殿
斋宫
奉先殿
大佛堂
寿康宫
慈宁宫
乾清门
隆宗门
景运门
保和殿
中和殿
太和殿
内务府
弘义阁
体仁阁
武英殿
武英门
太和门
金水桥
文渊阁
主敬殿
文华殿
文华门
断虹桥
十八槐
西华门
东华门
午门
西雁翅楼
东雁翅楼
角楼
角楼
筒子河
筒子河
倦勤斋
景祺阁
颐和轩
乐寿堂
养性殿
畅音阁
古华轩
宁寿宫
皇极殿
宁寿门
皇极门
九龙壁
箭亭
南三所
西所
中所
东所
影壁
三座门
国史馆书库
銮仪卫

紫禁城内廷外东路

The Outer Eastern Section of Inner Court of the Forbidden City

紫禁城内廷外东路主要包括宁寿宫一区和南三所、上驷院等。
宁寿宫一区格局仿照紫禁城中路，也有外朝与内廷之分，
是乾隆皇帝预为自己归政后的养老所。
南三所在明代为慈庆宫，是皇太子的生活区，
因为位居紫禁城东，所以有“东宫”之称。
清乾隆年间，南三所格局定型，
亦是众皇子的居所，所以又有“阿哥所”之称。
南三所东还有太医院、药王殿和御药房。
为皇帝圈养御马的上驷院和国史、
会典馆也设在内廷外东路。

锡庆门 / 1900年

Xiqing Men (Gate of Bestowal of Congratulations), 1900

锡庆门与敛禧门东西相对，是进入宁寿宫区的重要门座，门内为皇极门、九龙壁，门外可通御茶膳房、箭亭、景运门等处。

皇极殿 / 20世纪初

Huangji Dian (Hall of Imperial Supremacy), Early 20th Century

皇极殿是宁寿宫区规模最大的宫殿，仿照内廷中路乾清宫修建。

宁寿宫一区是乾隆帝在61岁时为自己86岁归政退位预为修建的养老所。这里仿照紫禁城中路，亦为外朝、内廷格局。嘉庆元年（1796年）元旦，乾隆帝在太和殿禅位后，初四日带领嘉庆帝在宁寿宫皇极殿举行了一场有三千余人参加的千叟宴，这是清宫历史上第四次，规模最大也是最后一次千叟宴。皇极殿见证了当年二帝同台、千叟交觥的宏大场面，繁盛至极，空前绝后。

皇极殿西北侧 / 1900年

Northwest side of Huangji Dian, 1900

光绪二十年（1894 年），慈禧太后六十寿辰时，将皇极殿原金龙和玺彩绘改为苏式彩绘。

皇极殿 / 1900年

Huangji Dian, 1900

皇极殿前原为新年竖立灯杆的须弥座上，被添置了一座六角重檐小亭。

皇极殿内景 / 20世纪初

Interior Scene of Huangji Dian, Early 20th Century

光绪二十年(1894 年),慈禧太后在皇极殿度过了六十岁寿辰;光绪三十年（1904 年），光绪帝在此接见了美国、奥匈帝国等九国公使。

乾隆帝晚年朝服像／故宫博物院藏

Qianlong Emperor in Court Dress, The Palace Museum

乾隆元年（1736年），登基伊始的乾隆帝曾向上天默祷，表示不愿超过祖父康熙帝六十一年的在位时间，如寿享86岁，在位六十年时即当归政，禅位嗣皇帝。嘉庆元年（1796年）元旦，他如愿将皇位授予皇十五子永琰，“心愿符初”地过上了太上皇的生活。其于乾隆三十七年（1772年）先期葺治的宁寿宫也就成为了太上皇颐养天年的场所。

宁寿宫 / 1900年

Ningshou Gong (Palace of Tranquility and Longevity), 1900

今天的宁寿宫原为宁寿宫后殿，皇极殿建成后改原宁寿宫为皇极殿，宁寿宫匾额移至后殿。乾隆皇帝在《宁寿宫铭》中说：“余将来归政时，自当移坤宁宫所奉之神位、神竿于宁寿宫，仍依现在祀神之礼。”说明了宁寿宫的改建是为了因循紫禁城中路的坤宁宫，在这里举行萨满祭祀礼。

养性门／1922年

Yangxing Men (Gate of Cultivation of Character), 1922

养性门形制与乾清门相同。门前“横街”将宁寿宫一区分为外朝、内廷两个区域。

养性门前鎏金铜狮 / 1924~1927年

A Golden Lacquer Bronze Lion in front of Yangxing Men, 1924-1927

养性殿院落 / 1922年

Courtyard of Yangxing Dian (Hall of Nature Cultivation), 1922

养性殿规制仿照养心殿，曾是乾隆帝预为自己归政后的寝宫。但自嘉庆元年（1796 年）元旦授玺，至四年（1799 年）正月初三驾崩，这三年零三天里，太上皇弘历一日也没有住过这里。

乐寿堂 / 1922年

Leshou Tang (Hall of Pleasure and Longevity), 1922

乐寿堂与养性殿一样，都曾是乾隆帝预为自己归政后的寝宫。慈禧太后六十大寿时，这里曾作为太后寝宫使用。乐寿堂的兴建参仿了长春园淳化轩，其奢华的内部装饰是现存乾隆时代宫殿内檐装修的代表。

乐寿堂明间青玉“丹台春晓图”玉山（寿山）／1925~1949年

A Shoushan Stone Carved with a Painting of Dantai Chunxiao in the Bright Room of Leshou Tang, 1925-1949

乐寿堂明间青玉云龙纹瓮（福海）/ 1925~1949年

A Green Jade Fuhai Carved with Clouds and Dragons in the Bright Room of Leshou Tang, 1925-1949

照片中乐寿堂西间门楣上悬挂着慈禧太后的肖像。门口还保存着清末写有“天下太平”字样的吉祥牌。

畅音阁戏楼 / 1925~1949年

The Peking Opera Theatre of Changyin Ge (Pavilion of Pleasant Sounds), 1925-1949

阅是楼内慈禧太后观戏处 / 1922年

A Luohan Bed for Empress Dowager Cixi to Watch Peking Opera in Yueshi Lou (Building for Seeing Opera), 1922

紫禁城畅音阁、圆明园清音阁、避暑山庄清音阁、颐和园德和园戏楼、紫禁城寿安宫戏楼是历史上清宫建造的五座三层大戏楼，保存至今的仅有畅音阁与德和园两座。畅音阁戏楼前建有二层阅视楼，是帝后观戏场所；左右有游廊，宗室大臣在此观戏；阁后又设扮戏楼，为参演人员准备之所。旧时，清宫太监为区分大小，曾唤避暑山庄清音阁为“大爷”，颐和园德和园戏楼为“二爷”，紫禁城畅音阁为“三爷”。

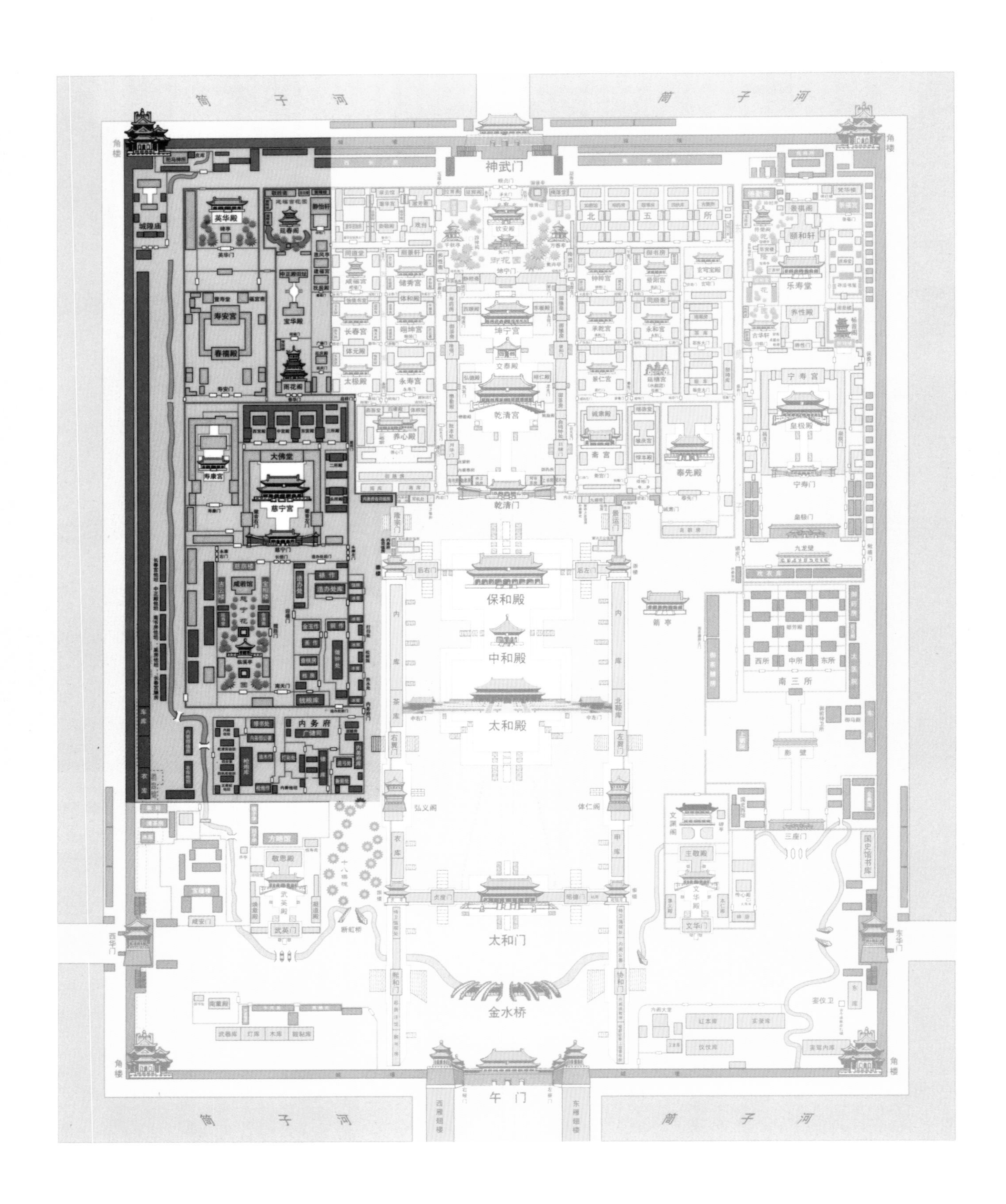
筒子河
筒子河
角楼
神武门
城隍庙
英华殿
寿安宫
春禧殿
雨花阁
宝华殿
北五所
御花园
钦安殿
储秀宫
翊坤宫
长春宫
永寿宫
太极殿
钟粹宫
景阳宫
承乾宫
永和宫
景仁宫
坤宁宫
乾清宫
养心殿
斋宫
奉先殿
乾清门
颐和轩
乐寿堂
养性殿
宁寿宫
皇极殿
宁寿门
皇极门
九龙壁
大佛堂
寿康宫
慈宁宫
慈宁门
慈宁花园
咸若馆
内务府
保和殿
中和殿
太和殿
南三所
弘文阁
体仁阁
文渊阁
主敬殿
文华殿
文华门
国史馆书库
方略馆
敬思殿
武英殿
武英门
断虹桥
太和门
熙和门
协和门
金水桥
西华门
东华门
南薰殿
午门
西雁翅楼
东雁翅楼

紫禁城内廷外西路

The Outer Western Section of Inner Court of the Forbidden City

紫禁城内廷外西路主要建筑包括慈宁宫、
慈宁花园、寿康宫、寿安宫、英华殿、雨花阁、
中正殿一区、建福宫花园等。
皇帝去世以后，慈宁宫、寿康宫、寿安宫是先朝后妃的养老所。
这里偏居一隅，较为封闭，
寡居高墙内的后妃们大多以念佛消磨时光，
所以外西路拥有多座佛堂。
中正殿一区是紫禁城中规模最大的宗教场所，
由雨花阁一直延伸到建福宫花园，
后者则是乾隆帝修建的一座精巧别致的宫殿园林。
此外，服务宫廷的内务府公署和众多“作处”亦在内廷外西路内。

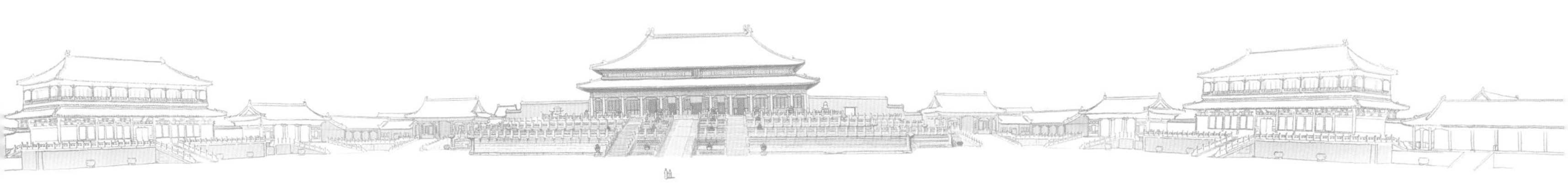

慈宁宫／20世纪初

Cining Gong (Palace of Compassion and Tranquility), Early 20th Century

慈宁宫内景／20世纪初

Interior Scene of Cining Gong, Early 20th Century

慈宁宫主要是为皇太后举行重大典礼的殿堂，凡遇皇太后圣寿节、上徽号、进册宝、公主下嫁，均在此处举行庆贺仪式。特别是皇太后寿辰时，皇帝亲自率众行礼，并与近支王公一同彩衣起舞，礼节十分隆重。若皇太后崩，梓宫奉安于慈宁宫中，皇帝至此行祭奠礼。

《胪欢荟景图》册之《慈宁燕喜》／故宫博物院藏

The Painting of Cining Yanxi from a series of paintings of Luhuan Huijing, The Palace Museum

上图描绘了崇庆皇太后万寿时，乾隆帝为其捧觞起舞的场景。此时的慈宁宫为单层檐，乾隆三十四年（1769 年），慈宁宫改建为今天的重檐形制。

慈宁花园临溪亭 / 20世纪初

The Linxi Ting (Pavilion over a Pound) of Cining Garden, Early 20th Century

慈宁花园临溪亭 / 20世纪初

The Linxi Ting of Cining Garden, Early 20th Century

慈宁花园是紫禁城内的四座花园之一，位于慈宁宫南侧，是一座轴线对称的宫殿园林。主要建筑包括：咸若馆、宝相楼、吉云楼、临溪亭、慈荫楼等。临溪亭居于花园中部，上亭下池，四角攒尖顶。旧时住在紫禁城外西路的太后、太妃们常到此消磨时光。

慈宁花园咸若馆 / 20世纪初

Xianruo Guan (Temple for Worshipping Buddha) of Cining Garden, Early 20th Century

慈宁花园慈荫楼 / 20世纪初

Ciyin Lou of Cining Garden, Early 20th Century

慈宁花园 / 20世纪初

Cining Garden, Early 20th Century

第一任故宫博物院院长易培基题临溪亭匾额

Tablet of Linxi Ting, Wrote by the first director of the Palace Museum Yi Peiji

慈宁花园宝相楼 / 20世纪初

The Baoxiang Lou (Building for Images of Buddhas) of Cining Garden, Early 20th Century

宝相楼与吉云楼相对，是慈宁花园咸若馆东配楼。乾隆时按照藏传佛教格鲁派的宗教法式，改建为二层七间楼阁。楼中依般若、功行、德行、瑜伽、无上阳体根本、无上阴体根本来布置供奉与陈设，此即“六品佛楼”的由来。宝相楼是紫禁城中现存的两座六品佛楼之一。

宝相楼内供奉的六座珐琅塔 / 20世纪初

Six Porcelain Enamel Pagodas, Early 20th Century

宝相楼一层明间供奉释迦牟尼佛，南北六间按照密宗法式供奉着六座珐琅塔。珐琅塔铸造于乾隆四十七年（1782年），六座形制各异，有楼阁式、覆钵式、密檐式、汉藏合璧式等，雕饰繁复，美轮美奂。

春禧殿东配殿
1925~1949年
The East Wing Hall of Chunxi Dian (Hall of Happiness of Spring), 1925-1949

春禧殿西配殿
1925~1949年
The West Wing Hall of Chunxi Dian, 1925-1949

寿安宫 / 1925~1949年

Shou'an Gong (Palace of Tranquil Old Age), 1925-1949

寿安宫明代称咸熙宫、咸安宫，雍正时为官学所在地。乾隆年间为庆贺崇庆皇太后万寿，改建为寿安宫，并在寿安宫前修建三层大戏楼一座，戏楼在嘉庆时拆除。寿安宫是清代太后、太妃、太嫔们居住的地方。康熙废太子允礽亦曾囚禁于此。此帧照片中，寿安宫门前张贴着“殿本书库”字样，说明拍摄时间在故宫博物院成立以后。

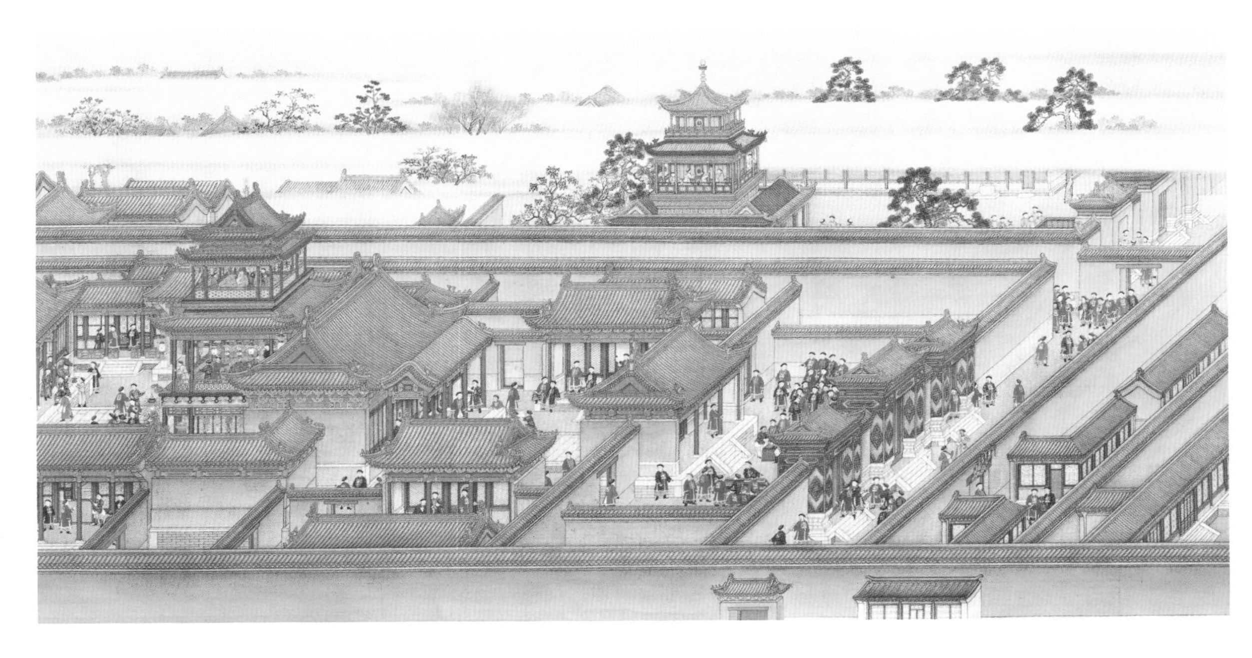

《崇庆皇太后万寿庆典图》卷中的寿安宫三层大戏楼 故宫博物院藏

The Three-Floor Opera Theatre of Shou'an Gong from The Painting of Empress Dowager Chongqing's Birthday, The Palace Museum

寿安宫配楼 / 1925~1949年

Wing Building of Shou'an Gong, 1925-1949

寿安宫西侧配楼 / 1925~1949年

West Wing Building of Shou'an Gong, 1925-1949

在乾隆年间，寿安宫院落的四周，围绕着三层戏楼建有一圈二层配楼。每到年节开戏时，寿安宫配楼是官员、宗室、外藩观戏的场所。这样配楼与戏楼相互搭配的建筑格局也出现在圆明园与避暑山庄的两座清音阁周围。可惜的是，圆明园同乐园清音阁、避暑山庄东宫清音阁与紫禁城寿安宫戏楼均已不存。

雨花阁四层佛龛供案 / 1922年

Four Tiers Shrine of Buddha in Yuhua Ge, 1922

雨花阁四层为无上瑜伽层，供案上供奉着密集金刚、大威德金刚、上乐金刚各一尊。

雨花阁 / 1922年

Yuhua Ge (The Rain Flower Pavilion), 1922

雨花阁是根据乾隆时代的国师三世章嘉呼图克图的建议，模仿西藏阿里古格托林寺坛城殿修建的一座密宗佛堂。雨花阁虽然外观三层，内部却严格按照藏密的事、行、瑜伽、无上瑜伽四部设计为四层的结构，每层分别供奉着藏密的四部神祇。雨花阁是目前我国现存最完整的藏密四部神殿，对于研究藏传佛教具有重要意义。

雨花阁 / 20世纪初

Yuhua Ge, Early 20th Century

由宝华殿望雨花阁 / 20世纪初

Yuhua Ge, from the angle of Baohua Dian, Early 20th Century

梵宗楼 / 1900年

Fanzong Lou, 1900

梵宗楼一层供奉文殊菩萨，二层供奉文殊菩萨的化身大威德金刚。在清代，由于文殊菩萨的译音“曼殊师利”与“满洲”发音相近，所以达赖与班禅两活佛在奏书中常常称皇帝为“曼殊师利大皇帝”，由此，清代帝王便成为了文殊菩萨的化身。梵宗楼内主供文殊菩萨，所以相应地这里存放着其化身乾隆帝的盔甲、刀枪弓箭、兽皮等物。

雨花阁北面 / 1900年

North Side of Yuhua Ge, 1900

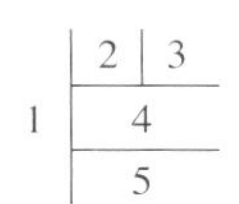

1.宝华殿前铜香炉 / 1900年

A Xiang Lu in front of Baohua Dian, 1900

宝华殿前“三足宝鼎青铜大香炉”，立于汉白玉须弥座上，炉上铸有铭文“大清乾隆乙巳年造”。

2.宝华殿内景 / 1900年

Interior Scene of Baohua Dian, 1900

宝华殿是清宫中正殿佛堂区中主供释迦牟尼佛的一处佛堂。清代，宝华殿明间设四方铜镀金大龛一座，内供金胎释迦牟尼佛一尊。龛前供案上供观音菩萨和阿弥陀佛铜像。东、西次间沿墙供案上亦陈设佛像、供器。这里的日常佛事活动主要是喇嘛诵经和设供献等。清代皇帝每年数次到这里拈香行礼。

3.宝华殿法器 / 1900年

The Ritual Implements of Baohua Dian, 1900

4.中正殿内供案 / 1900年

The Altar in Zhongzheng Dian, 1900

中正殿是建福宫南侧一座主供无量寿佛的殿宇，清宫常把以中正殿为中心的一组佛堂建筑（中正殿、中正殿后殿、东西配殿、香云亭、宝华殿、雨花阁、雨花阁东西配楼、梵宗楼十处）总称为“中正殿”。

5.“扮鬼”的喇嘛 / 1920~1923年

Lamas with Ghost Masks, 1920-1923

照片上注：每年二月初一日，喇嘛戴面具演习“打鬼”情形。所谓“打鬼”，即藏传佛教的“跳步扎”，是一种驱邪去祟的仪式。据《清会典》记载，“打鬼”时“布达拉众喇嘛，装诸天神佛及二十八宿像，旋转诵经。又为人皮形，铺天井中央，神鹿五鬼及护法大神往捉之。末则排兵甲幢幡，用火枪送至布达山，以除一岁之邪。”

建福宫花园延春阁 / 1920~1923年

Yanchun Ge (Pavilion of Prolonged Spring) of Jianfu Gong (Palace of Establishing Happiness) Garden, 1920-1923

建福宫花园又称西花园，是紫禁城中的四座花园之一。这里殿阁亭馆错落有致，假山叠石点缀其间，是一座集庄严大气与玲珑秀美于一身的宫殿园林。1923 年 6 月，建福宫花园的绝大部分建筑被火焚毁。

由延春阁望北海琼华岛 / 1920~1923年

Qionghua Dao (Jade Flower Island) of Beihai, from the angle of Yanchun Ge, 1920-1923

照片忠实地记录下大火前建福宫花园西侧一隅的景象。近景卷棚歇山小殿为碧琳馆，北侧即模仿圆明园半亩园的敬胜斋，两座建筑均被焚毁。

1923 年 6 月 26 日晚 9 时，建福宫花园敬胜斋火起，大火很快延烧到南邻的延春阁，高大的延春阁掉落的构件又引燃了周围殿座。由于建福宫花园地处内廷北侧，建筑密度大，附近又没有可以利用的水源，在很短时间内花园的中路和西路就陷入了火海，就连毗邻的中正殿一区也未能幸免。至 28 日凌晨，意大利使馆派 30 余名士兵参与扑救，大火才最终被扑灭。据事后统计，大火烧毁了敬胜斋、静怡轩、延春阁、吉云楼、慧耀楼、碧琳馆、妙莲花室、积翠亭、玉壶冰、广生楼、香云亭、中正殿等建筑上百间，贮存其中的大量珍玩、典籍、宗教法器被焚毁，损失不可胜计。

建福宫花园
积翠亭与广生楼
1920~1923年
Jicui Ting and Guangsheng Lou of Jianfu Gong Garden, 1920-1923

建福宫花园
存性门火后残迹
1923年6月
The Relic of Cunxing Men, after the fire of Jianfu Gong Garden, June 1923

建福宫花园延春阁火后残迹
1923年6月
The Relic of Yanchun Ge, after the fire of Jianfu Gong Garden, June 1923

建福宫花园虎皮墙圆光门火后残迹
1923年6月
The Relic of Hupi Wall and Yuanguang Men, after the fire of Jianfu Gong Garden, June 1923

下篇
皇家苑囿与陵寝

满族统治者定都北京以后，对京城宫苑的修整与改造被提上了日程。这一时期主要任务是将李自成起义军留下的破坏和明末以来因年久失修造成的建筑损毁修缮一新，以适应新朝气象；此时在宫苑中创新了包括萨满祭祀场所在内的一些符合满族统治者民族习惯的建筑，增建了诸如琼华岛白塔等一批藏传佛教建筑，以迎合统治者的宗教信仰；在城内大行满汉之别，将北京城划分在八旗驻防的分区管领之下，为数众多的汉族原住民被迫迁向外城……经过短暂修整与割划后的帝都北京，以一种前所未有的面貌迎奉着她的新主人。

北京城内的拥挤不堪与四季分明的时令特色令满族统治者对这座古老又崭新的帝都

大呼不适。在炎炎盛夏，从紫禁城到西苑三海，再到南苑猎场，甚至远至喀喇河屯的“避暑边城”，皇帝带领大臣们不断地外出巡幸，时间动辄月余。他们对曾经关外重峦叠嶂、山水秀美的自然环境继续抱持着一种眷恋不舍的感情。这促使他们在京城内外寻找适合的地方，建立能够分担统治功能的郊野离宫。这就是清代皇家大兴园林建设的最直接的动因。所以，诸如西苑三海、三山五园、避暑山庄、静寄山庄等皇家园林在短短一百年的时间里相继建成，特别是以三山五园为代表的海淀皇家园林群，三百年来享誉世界。

清代，海淀园林从起步发展到日臻全盛，是有着自然与社会历史双方面原因的。就自然原因来说，“水乡”是涵养园林的摇篮。海淀的自然地理环境为园林建设准备了必不可少的硬件条件，这里的山水形胜构成了建造园林的骨架。就社会历史原因来说，清代的康熙、雍正、乾隆三朝是中国封建时代最后的辉煌期，社会秩序相对稳定，国家有充足的财力经营园林。更重要的是，三帝均有较高的文化修养，他们本身就扮演着园林设计者的角色。从规划布局到建设施工，每一座敕建园林都将自然风致与人文情怀交融在一起，在皇帝的品题下，园林品位得到不断的提升。

上之所好，下必甚焉。在皇帝的率作下，士大夫阶层纷纷效法，私人宅邸的规模与数量也与日俱增。乾嘉之际，皇家园林、赐园、士绅宅园无论在规模还是数量上，都达到了历史上的全盛。尤其是赐园与士绅宅园，在皇家园林建设甚嚣尘上的影响下得到飞速发展。以致在圆明园周围，已无多少空地供官宦士绅们开辟私园。这从一个侧面解释了乾隆后期以来，皇家赐园的频率日渐增多，规模宏大的私人宅园却在日益减少的原因。

乾隆后期，包括西苑三海、三山五园在内的众多皇家园林与赐园，给清廷带来了巨大的财政负担。大小几十座园林无论官私，木构建筑的日常维护、水道河网的定期疏浚以及大量园户、匠役、护军的差饷等等，每年消耗的帑银达数十万两之巨，即便在国力全盛的乾隆时代，应付这样的花销也显得捉襟见肘。乾隆帝曾明发谕旨，谕令后世子孙切勿大兴土木，并一再强调，现有园林足够皇家享受，不可另择新址再建新园。乾隆帝甚至借题专作《知过论》告诫后世“勿兴作”、“惜民力”的道理。口实虽然冠冕堂皇，其实质与日益尾大不掉的工程虚耗有着莫大关系。皇帝的支持曾经是推动皇家园林走向繁荣的重要原因，但在国力江河日下的现实面前，不得不大为收敛甚至是有意缩减。例如，嘉庆帝增葺绮春园时，每年只动工一二处，御园中的年节花销也一再削减。道光帝更是索性“裁撤三山陈设”，将除去圆明三园、畅春园以外的其他西郊园林一概封存，彻底停罢了三山游幸。

1860年，英法两国悍然发动了第二次鸦片战争，被太平天国运动拖入十年战争泥潭的清政府无力抵御联军的进犯，终至咸丰帝北逃热河，英法联军兵临北京城下。他们是继嘉庆二十一年（1816年）英国阿美士德使团后首次进抵北京的西方人。这一次，洋人带来的不再是贡品与表文，而是快枪和大炮。10月6日，法军8000人在追击散逃的清军途中，得到指引而直抵清帝临朝视政的御园——圆明园，海淀园林的末日已经不可避免地到来了。10月6日下午，法军到达海淀，军官随即下令放火焚烧，百年之久的海淀镇街道市肆与园林宅邸迅即化为灰烬。10月6日黄昏，法军先头部队抵达圆明园大宫门外，在与二十几名技勇太监的短暂交火后，迅速占领了该园。一部分士兵即在此时开始了劫掠与破坏。10月7日，12000名英军陆续抵达圆明园，驻扎在大宫门外御路两侧。以英方全权特使额尔金与将军格兰特和法方全权特使葛罗与将军蒙托邦为代表的联军统帅部下达了对以圆明园为中心，包括周围几十座附属园林在内的整个海淀园林群的抢掠命令。在三天的时间里，双方士兵按单位、军种分批进园抢掠。各种肤色、操持不同语言的士兵争先恐后在园中大肆劫掠，他们役使中国人搬运琳琅满目的“战利品”，动用枪械任意损毁一切无法带走的物品。上古鼎彝、晋唐碑帖、宋元文存及至《四库全书》等重要文化典籍悉数遭劫。清廷积数百年之力贮藏于园中的珍贵收藏几乎被抢劫殆尽。在以圆明园为中心，方圆10公里的范围内，三山五园及其附属园林、士绅宅邸多被袭扰，各园殿座均有破坏。10月18日，在额尔金与格兰特的指令下，英军米启尔骑兵团3000名士兵手持火把放火烧园，并再一次洗劫了前次未至之处。三山五园的园林建筑在两天不熄的大火中所存不足十之二三，只有清漪园前山与静明园山区的零星建筑免遭灭顶之灾，但其内部陈设早已被联军洗劫一空。如此景象，记录下了人类文明史上最为惨烈的一页。

1900 年 6 月，八国联军进占北京，刚刚修复不久的颐和园再次被侵略军占领，虽然没有发生 40 年前大规模焚烧园林的事件，但清末以来积聚园中的宫廷陈设大部分被掠去。在秩序大乱中，海淀地区 1860 年以来残存的园林宅邸又大多被不法之徒抢掠、破坏。除颐和园与后来新建的部分宅园外，其他传承自清中期的园林宅邸再也没有被恢复起来。

在本篇中，编者挑选了一批涉及西苑三海、三山五园、避暑山庄等具有代表性的皇家园林照片，拍摄时间从1873年到20世纪40年代，跨越了清末到民国的六七十年。这段时间，正是清代皇家园林从极盛而衰到间歇性恢复的过程。从照片中我们既可以领略遗留自乾隆时代的踵事增华，又可以窥见清末皇家重拾破碎园林的努力，这些照片为我们审视历史，提供了绝佳的素材。

除了皇家苑囿，本部分还收录了一批皇家陵寝的照片。例如清帝在关内的第二座皇家陵区——清西陵。从雍正帝始，清代统治者们开始在京南易县永宁山下建造万年吉地，形成了关内东西二陵并存的局面。乾隆皇帝还特别制定了父子东西埋葬的“昭穆之制”，两座陵区得以香火并盛，一脉相传。值得一提的是，本篇收录的清西陵照片中，有 40 余帧是拍摄于 1909~1913 年之间，光绪帝崇陵兴建过程的施工纪实照，从测量、施工到工程完竣，都留下了纪实照片。这些照片是内务府司事人员向皇帝汇报施工过程与竣工情况的“呈览”本，为研究清代大型皇家工程各个阶段的施工程序与做法提供了珍贵的第一手资料。本篇另录有农事试验场、黑龙潭、乐净山斋、醇亲王园寝等处照片。乐净山斋虽然与皇家没有太直接的关联，但也因其主人身份的特殊而划入皇家建筑之列。

本篇所涉及的苑囿与陵寝建筑，或仍然保存至今，或早已沦为残迹，读者朋友们可以凭借着这些照片来对比古今异同、凭吊盛衰，感受时代的变迁。

王志伟

故宫出版社宫廷历史编辑室

Imperial Gardens and Mausoleums

After Manchu rulers declared as the "Son of Heaven" in Beijing, the project of repairing and reconstructing imperial buildings was putting on agenda. In order to proclaim "Qing" as the new ruler of entire China, the main target of the project aimed to firstly reconstruct imperial buildings, which were damaged by the rebel forces led by Li Zicheng during the late Ming dynasty. Therefore, large portions of Palaces (e.g. Kunning Gong) converted for Shamanist worship of Manchu rulers; the new buildings of Tibetan Buddhist worship like Bai Ta of Qionghua Dao (White Pagoda of Jade Flower Island) were built in the Sea Palace; and the new division of inner city in terms of the Eight Banners under the new sovereign of Qing dynasty, which caused most of the local citizens of Han had to move out the inner city of Beijing.

One of the most significant changes of Beijing after the settlement of Manchu ruler, it was that the constructing of imperial gardens. The emperors spent most of their time in the Sea Palace, the Hunting Resort in Nanyuan and even went to the Summer Resort in Kala Hetun for summer. It was because that the Manchu ruler had a difficult time to adjust to the climate and overcrowded population of Beijing and overcome with nostalgia of their time outside the Shanhai Pass. Therefore, several beautiful imperial gardens were built during the sovereign of Qing dynasty, such as the Sea Palace, Three Mountains and Five Gardens, Imperial Summer Resort and Jingji Resort. All these magnificent imperial gardens were built only over one hundred years and the most significant ones were Haidian Gardens, which were represented by the Three Mountains and Five Gardens.

There are two reasons in terms of the achievements of Haidian Gardens in Qing dynasty: nature and social history. Talking about nature, water is one of the most fundamental factors on building gardens. The geographical location of Haidian provides a perfect natural location to imperial gardens. Secondly, in terms of social history, the reign of Kangxi Emperor, Yongzheng Emperor and Qianlong Emperor was the last consolidation period in ancient Chinese history; the government of Qing had more financial budget to build these gardens. Furthermore, these three emperors were all well educated and had the knowledge on culture and fine art. They also were the designers of these imperial gardens. Under the supervision of emperors, the imperial gardens were integrated with natural beauty and cultural arts from planning to constructing.

Due to the emperors' favor of gardens, the courtiers started to build gardens as well. Numerous private gardens were built during the reign of Qianlong Emperor and Jiaqing Emperor. The number and the measurable area of the gardens had reached a significant percentage in ancient Chinese history. Among all these gardens, a large number was taken by the granted gardens and courtiers' gardens. Consequently, there was no more spare room for people to build private gardens near the Yuanming Yuan (Old Summer Palace). This phenomenon might explain the reason why the numbers of granted gardens were increasing, while the numbers of private gardens were decreasing during the late reign of Qianlong Emperor.

However, the cost of maintaining imperial gardens was a huge financial burden to the Qing government during the late reign of Qianlong Emperor. The daily maintaining, including the payment of gardeners and guardians, of imperial gardens would cost more than 100,000 silver coins (approx. 4.2 millions dollars) every year. Even in the reign of Qianlong Emperor, the Qing court could hardly cover the cost of imperial gardens. From the imperial rescript of Qianlong Emperor, he emphasized that there was no need to build new imperial gardens. Meanwhile, he also wrote an article of *Reflections of My Own Fault*, to warn his successors that building imperial gardens would put a huge burden on citizens and it was not a right thing for emperors to do. The support from emperors used to be the most important factor on building imperial gardens, but the financial crisis could not cover the huge cost of the gardens. Therefore, the successors after Qianlong Emperor started to reduce the budget on imperial gardens. For example, while repairing the Qichun Yuan (Elegant Spring Garden), Jiaqing Emperor ordered the workers that they only allowed to repair one or two sites of the garden in order to save budget. Daoguang Emperor even closed all the western imperial gardens except three gardens of Yuanmin, and Changchun Yuan (Garden of Exhilarating Spring).

In 1860, the outbreak of the Second Opium War was pitting British Empire and the second French Empire against Qing dynasty of China. With the Qing army devastated, Xianfeng Emperor fled capital to Rehe (Jehol). On the night of 6th October, 8000 French soldiers diverted from the main attack force towards the Yuanming Yuan. Only a few eunuchs tried to fight against the French army but failed, later the palace was occupied by the French units. On 7th October, 12,000 British soldiers arrived on the Yuanmin Yuan. Under the command of the British and French commander, there was extensive looting by Anglo-French army. After the looting, on 18th October, Lord Elgin ordered the destruction of the palace. It took 3000 British troops to set fire on the entire palace, taking a total two days to burn. Only three of ten buildings survived, including the front mountain of Qingyi Yuan (Garden of Limpid Ripples) and the mountain area of Jingming Yuan (Garden of Brightness with Quiescence).

In 1900 during the Eight-Nation Alliance invasion, Yuanming Yuan was again looted and destroyed. Even though the invaders did not burn down the palace like the Anglo-French troops, the treasures and imperial buildings were looted and completely destructed. Furthermore, most of the imperial gardens from mid-Qing dynasty, which survived during the destruction of 1860, were all destroyed except Yihe Yuan (Summer Palace) and some new buildings that were built after the first destruction.

In this section, we choose a series of photographs of the Sea Palace, Three Mountains and Five Gardens and Imperial Summer Resort. It covers almost 7 decades from 1873 to 1940s, featuring the period of late Qing dynasty and Republic of China. From the collection, it provides a vertical timeline of the history of imperial gardens: the glorious time of the reign of Qianlong Emperor and the destruction during the Second Opium War and Eight-Nations Alliance invasion. The section also demonstrates projects that how the workers repair the imperial gardens.

Besides the imperial gardens, the section also comprises the photos of imperial mausoleums such as West Mausoleum of Qing dynasty. Since the reign of Yongzheng Emperor, the Manchu rulers started to establish their mausoleums at southwest of Beijing in Yi County. Qianlong Emperor decided that he should be buried in the Eastern Qing tombs and dictated that thereafter burials should alternate between the eastern and western sites. In this section, there are 40 photographs, which were taken between 1909 and 1913, documented the project that building the Chongling (Mausoleum of Guangxu Emperor), including the construction and measurement of the mausoleum. The bondservants of Neiwu Fu (Imperial Household Department) captured all these historical photos in order to report the procedure on building the mausoleum.

The section also comprises photos of Experimental Farm (late transferred to Beijing Zoo), Heilong Spring, Lejing Shanzhai and Tomb of Prince Chun. Although Lejing Shanzhai and Tomb of Prince Chun do not have direct connection of imperial court, they still belong to the imperial buildings in terms of their owner's noble status. These photos are the documental photographic collection of the Palace Museum.

Some of the imperial gardens and mausoleums in this section still exist in nowadays, some may have already become into relics. We hope our readers could feel the changes of time through this photographic collection, by comparing the past with the current days.

Wang Zhiwei
The Forbidden City Publishing House

景山

Jingshan

景山，又称煤山、万岁山，成山于金元时期。明代营建紫禁城时，
用折除元朝宫殿的渣土和挖掘护城河河道土方堆积在山上，
清代始称景山。景山五峰上建五亭，分别是万春亭、观妙亭、
辑芳亭、周赏亭、富览亭，五亭内供奉着五方佛。
景山北侧，建有一组规模宏大的建筑群——寿皇殿，
是清代帝王奉祀先帝御容与神主的皇家祖庙。
寿皇殿东侧的永思殿又是清帝为先帝停灵、哭奠的场所。
整座景山，山林苍翠，殿阁尊崇，环境肃穆，处处体现着统治者对祖先的虔敬。
站在北京城中轴线的制高点万春亭上眺览四方，
“南瞻百阁千丹阙，北俯万巷亿蓬瀛”，帝都的壮美，无以逾此。

景山与东三座门 / 1901年

Jingshan and the East Three Gates, 1901

此帧照片以全景的视角将景山与东三座门收入镜头中。门内护城河北岸，是一排东西连通的长房，在北上门和东西三座门拆除以前，这里是一处相对封闭的区域，旧时的景山官学设在此处。

景山万春亭 / 1925~1949年

Wanchun Ting (Pavilion of Ten-Thousands-Spring) of Jingshan, 1925-1949

景山中峰高 45.7 米，万春亭建在中峰之巅，是一座三重檐五开间方亭，亭内原供五方佛之首的毗卢遮那佛。1900 年，八国联军进占紫禁城和景山西苑，五亭中的五方佛在此时下落不明。

景山观妙亭 / 1925~1949年

Guanmiao Ting of Jingshan, 1925-1949

景山辑芳亭 / 1925~1949年

Jifang Ting of Jingshan, 1925-1949

照片近景是辑芳亭，远景可见大高玄殿院内乾元阁、陟山门街、御史衙门、琼华岛白塔、妙应寺白塔等，大西山也若隐若现在背景中。

景山辑芳亭 / 1925~1949年

Jifang Ting of Jingshan, 1925-1949

景山辑芳亭 / 1922年

Jifang Ting of Jingshan, 1922

两帧照片均可见紫禁城东北角楼、大高玄殿牌楼和习礼亭。护城河边的长房在上一帧照片中已经不见踪影。

寿皇门前西侧牌坊／20世纪初

Paifang at the West Side of Shouhuang Men, Early 20th Century

寿皇门外东、西、南三面设坊，围合出相对封闭、静穆的空间。牌坊木质，四柱三间九楼。坊额“旧典时式”为乾隆帝御书。

寿皇门前东侧牌坊与石狮 / 20世纪初

Paifang and Stone Lions at the East Side of Shouhuang Men, Early 20th Century

寿皇殿 / 20世纪初

Shouhuang Dian, Early 20th Century

寿皇殿旧址在景山东北，乾隆十四年（1749 年）在景山子午线北侧新建寿皇殿。寿皇殿明堂九室，重檐庑殿。另有寿皇门、东西配殿、井亭、碑亭、神厨、神库等附属建筑，整体规制与圆明园安佑宫略同，是清帝继太庙、奉先殿、安佑宫后的又一组奉祀先祖的建筑。乾隆时，先于寿皇殿内供奉康熙、雍正二帝御容，后将紫禁城体仁阁中存放的努尔哈赤、皇太极、顺治诸帝后御容移至寿皇殿左侧的衍庆殿内保存，犹如祧庙之制。

寿皇殿与丹陛石 / 约1907年

Shouhuang Dian and Danbi Shi, Approx. 1907

寿皇殿内景 / 20世纪初

Interior Scene of Shouhuang Dian, Early 20th Century

寿皇殿内九间九室，前设供案、灯檠，分别供奉清代历朝帝后神像。这种一殿分数室的供奉布局，称为“同堂异室”，是祖庙建制的形式之一。

景山北望 / 20世纪初

The View from the North Side of Jingshan, Early 20th Century

此帧照片的近景是寿皇殿建筑群，远景为皇城北门地安门、万宁桥、鼓楼、钟楼。一条南北贯通的北京城中轴线清晰可见。照片隐约可见北京内城北侧城墙。

西苑三海

Xiyuan Sanhai (Sea Palace)

在北京皇城以内，南北分布着三片开阔的水域——北海、中海、南海。

因其建在皇宫的西侧，所以被称为“西苑三海”。

辽金以来，帝王们在这里挖湖堆山，修造园林，作为政务之余的离宫别苑。

辽太宗葺筑“瑶屿行宫”，一开西苑宫殿园林的先河；

金统治者仿宋汴梁艮岳园，移花石纲旧物，筑广寒、瑶光二殿；

忽必烈三次扩建琼华岛，称“万岁山”，作为其朝会之所；

明代增建了团城和三海东、西、北三岸的殿宇楼台。

清代是西苑三海建设的繁盛期，顺治帝在广寒殿废址建喇嘛塔，

成为了此地三百年来的标志性景观。

乾隆时，曾对西苑进行了长达三十年的改扩建，

使这片距离紫禁城最近的皇家园林焕发了盛世的光彩。

由中海北望琼华岛／清末

Qionghua Dao (Jade Flower Island), from the angle of Zhonghai, Late Qing dynasty

此帧照片的拍摄者站在万善殿一侧岸边，将镜头对准北侧的琼华岛，岛上永安寺白塔、善因殿、庆霄楼清晰可辨。稍近一些的团城被全景摄下，西苑北侧宫墙一直延伸到九孔金鳌玉蝀桥前。透过大桥，照片清晰地表现出北海北岸几组宗教建筑群的风貌：五龙亭北侧，阐福寺山门后的三层大殿尚称完好地屹立在原址，1919 年大殿被火焚毁；左侧，小西天极乐世界正在揭瓦维修中；极乐世界后有万佛楼，楼内供奉着崇庆皇太后六旬万寿时，王公大臣进献的 10298 尊佛像，遂得“万佛楼”之名，这座建筑在 20 世纪六七十年代被陆续拆除。

从积翠坊望琼华岛 / 1915~1920年

Qionghua Dao, from the angle of Jicui Fang, 1915-1920

金鳌玉蝀桥 / 20世纪初

Jin'ao Yudong Qiao (Bridge of Golden Sea Monster and Jade Rainbow Trout), Early 20th Century

北海团城承光殿内玉佛 / 20世纪初

Jade Buddha of Chengguang Dian (Hall of Received Light) in Tuan Cheng (The Round City), Early 20th Century

团城承光殿中供奉释迦牟尼佛坐像，高1.5米，为整块白玉雕琢而成。

北海善因殿／1915~1920年

Shanyin Dian of Beihai, 1915-1920

善因殿建在白塔下，天圆地方形制，周身塑贴琉璃佛像。其内部精美的曼陀罗天花彩绘下，供奉着一尊大威德金刚。

北海西天梵境前琉璃牌坊 / 20世纪初

A Glazed Paifang in front of Xitian Fanjing (Hall of Western Paradise), Early 20th Century

西天梵境又称大西天，在北海北岸。明代时为经厂，乾隆二十四年（1759年）扩建。院内建有大慈真如宝殿（楠木殿）、七佛塔碑亭和大琉璃宝殿（无梁殿）等。此帧为着色照片。

北海九龙壁 / 1915~1920年

The Nine-Dragon Wall of Beihai, 1915-1920

从北海五龙亭望琼华岛 / 1915~1920年

Qionghua Dao, from the angle of Wulong Ting (The Five-Dragon Pavilion), 1915-1920

北海静心斋 / 约1907年

Jingxin Zhai (Lodge of Quiet Heart), Approx. 1907

中海紫光阁／1900年

Ziguang Ge (Palace of Purple Glory), 1900

紫光阁在中海西岸，是清帝校阅八旗子弟骑射和殿试武举之处。乾隆二十五年（1760年），为纪念平定准噶尔回部之功，乾隆帝重修紫光阁，将战功卓著的一百名功臣画像收藏在这里。此后，凡乾隆帝历次“十全武功”，皆绘功臣像“图形紫光阁”。终乾隆一朝，紫光阁内贮存的功臣像达200幅以上。除了功臣像外，历次战役中的灵纛、缴获敌军武器等均收藏在紫光阁中，以示纪念。每当战争得胜，清帝为酬劳凯旋将士，在紫光阁前搭起蒙古大幄，在这里举行规模盛大的凯旋宴。

1900年，西苑三海被八国联军占领，紫光阁内陈设文物悉数被掠。

乾隆帝《紫光阁赐宴图》卷（局部）
故宫博物院藏

A Dinner Granted by Qianlong Emperor at Ziguang Ge [Part], The Palace Museum

中海紫光阁 / 1922年

Ziguang Ge, 1922

紫光阁明间内景 / 1922年

Interior Scene of the Bright Room in Ziguang Ge, 1922

紫光阁内曾立有一通卧碑，上镌乾隆皇帝的两道谕旨，告诫八旗子弟不要耽于安乐而忘记国语骑射的传统。新中国成立后，这通卧碑被移至紫光阁后武成殿前，周恩来总理了解了碑文后，将其概括为“下马必亡碑”。

中海万善殿／20世纪初

Wanshan Dian (Hall of Innumerable Virtues) of Zhonghai, Early 20th Century

万善殿在中海东岸，靠近水云榭，重檐黑瓦，方广三间，是在明代蕉园崇智殿的基础上改建而来。殿内曾悬有顺治帝御书“敬佛”二字，供奉着三世佛及诸神。万善殿北，建有一座圆顶佛殿，名千圣殿，内供紫檀七级千佛塔一座。

中海水云榭 / 约1907年

Shuiyun Xie (The Water Cloud Kiosk) of Zhonghai, Approx. 1907

水云榭位于靠近中海东岸的湖面上，内有乾隆帝御笔“太液秋风”碑，此处是著名的燕京八景之一。

澂見商飈蘋末生鏡
瀾玉練影中橫非
關細雨頻傳響何事
平流自有聲爽入
金行閶闔表波連
瑤渚趯臺瀛高秋
文宴傳嘉話已覺
犁然今昔情
右太液秋風

张若澄绘、乾隆御题《燕京八景图》册之太液秋风 / 故宫博物院藏

Taiye Qiufeng from the Poetry “Eigth Scenes of Peking”, Wrote by Qianlong Emperor, Drew by Zhang Ruocheng, The Palace Museum

中海海晏堂宫门／20世纪初

The Entering Gate of Haiyan Tang, Early 20th Century

中海海晏堂建于光绪三十年（1904 年），是在原仪鸾殿旧址上仿长春园海晏堂建起的一座西洋楼。楼成后，这里作为慈禧太后接见外国公使夫人和女宾的场所。海晏堂宫门门额御书“寿箕翼”三字，门上的西洋蕃草装饰虽然极尽繁复，但缺乏节奏、变化与动感，少了长春园海晏堂西洋雕饰的灵动与变幻。尽管如此，此帧照片仍然为我们展现了巴洛克风格在清代宫廷建筑中的最后辉煌。

中海海晏堂／1912年

Haiyan Tang (Hall of National Peace) of Zhonghai, 1912

中海海晏堂仿照长春园海晏堂建造，虽然体量超过了后者，但在局部装饰与楼前水法的设置上还是逊色于长春园海晏堂。照片中海晏堂楼前水池两侧设有“十二生肖”铜像，与长春园海晏堂不同的是，这里的生肖铜像并不喷水，但它们仍然具备授时功能。奥妙就在铜像的手中各执一盏电灯，电灯会根据一天中时间的不同分别亮起。从长春园“水力钟”到中海“电力钟”，两座海晏堂水法的变迁体现了授时技术的进步。

1912年，袁世凯就任临时大总统时，邀孙中山来京会面，曾在海晏堂举行盛大宴会。第二年，袁世凯正式就任中华民国大总统后，改海晏堂为居仁堂，作为他办公、会客的场所，时称“总统楼”。1964年，居仁堂被拆除。

中海海晏堂明间内景／1912年

Interior Scene of the Bright Room in Haiyan Tang, 1912

据清宫档案记载，海晏堂落成后，其室内装修采用了中西结合的风格。明间设地平、宝座、屏风和相应的中式陈设；楼梯则采用西洋样式。室内悬挂着西式吊灯、挂屏，大量使用玻璃制品，并安设有多座自鸣钟表。从中海海晏堂的室内陈设可以想见长春园海晏堂的内檐风格。照片中的御座、御案被一张明黄色绸布罩住，以示对皇帝的尊重。

长春园海晏堂铜版画
故宫博物院藏

An Intaglio Painting on Haiyan Tang of Changchun Yuan, The Palace Museum

南海瀛台 / 1917~1924年

Ying Tai (The Ocean Terrace) of Nan Hai, 1917-1924

瀛台，在南海湖心，明代称“趯台坡”、“南台”。其上原有建筑名昭和殿，是明代帝王宴饮游幸之处。清代顺治、康熙、乾隆年间，对瀛台均有修葺，照片中的瀛台建筑多是乾隆时期留下的。瀛台的平面近似圆形，主要建筑轴线对称。从南向北依次是迎薰亭、蓬莱阁、涵元殿、涵元门、翔鸾阁，轴线两侧有长春、补桐两座书屋，另有怀抱爽、八音克谐、待月轩、牣鱼亭诸胜。

从新华门进入中南海，向北望去，水面上一片金碧楼台，四面波光烟霭，仿佛传说中蓬瀛三岛的境界。

南海翔鸾阁 / 1912年

Xiangluan Ge (Pavilion of Great Fragrance) of Nan Hai, 1912

翔鸾阁是一座二层五间楼阁，阁下开门，通向北侧的瀛台桥，它是瀛台中轴线上最后的建筑。旧时，翔鸾阁二层的东西梢间各安设一架直径超过一米的大自鸣钟，今已不存。

翔鸾阁叠落廊 / 1900年

Dieluo Lang of Xiangluan Ge, 1900

南海日知阁／1915~1920年

Rizhi Ge of Nanhai, 1915-1920

南海日知阁在淑清苑，与流水音东西相对，阁下为三海出水口，太液池水出日知阁后过织女桥，流入金水河。日知阁曾是乾隆帝十三四岁时读书的地方。乾隆帝即位后，将其皇子时写就的诗文辑成《乐善堂集》三十卷，“《乐善堂集》集中未备载者，复择其精要语二百六十余条别为四卷，名曰《日知荟说》。”《日知荟说》（见右图）定名的由来，即乾隆初学时代的日知阁。照片所示日知阁建于叠石水口上，形作一道卧虹，在明清宫廷建筑中极为少见。

《日知荟说》书影 / 故宫博物院藏

A Photo of Rizhi Huishuo, The Palace Museum

南海流水音 / 1900年

Liushui Yin of Nanhai, 1900

南海流水音在淑清苑，是一座四方流杯亭。康熙皇帝御书并刻石“曲涧浮花”，乾隆御题“流水音”。流水音与紫禁城乾隆花园禊赏亭功能相同，均模仿会稽山阴的兰亭修禊。亭内以石铺地，引水流入如意形水道，水道上可浮酒杯。旧时，清帝常带翰林词臣到此临流雅集，吟诗赓和。

南海宝月楼 / 20世纪初

Baoyue Lou (House of Precious Moon) of Nanhai, Early 20th Century

宝月楼在南海南岸，北与瀛台迎薰亭隔水相望。此处原无建筑，乾隆帝因南海南岸长堤一线，太过空旷，又狭长无屏蔽，遂下令建造此楼。站在宝月楼上，北顾浩森波光的瀛台仙境，南望回回营的异域风情，所以乾隆帝题下了“仰视俯察”的匾额。袁世凯当政时，宝月楼被改成中南海南门——新华门。宝月楼外回回营建于乾隆二十五年（1760 年），这里曾是清政府平定回部后迁居大小和卓族人的地方。为了尊重迁居维吾尔族人众的宗教习惯，乾隆帝特在西苑外建造一座清真寺，名“敕建回人礼拜寺”。寺成后，其宣礼楼（见右图）与西苑宝月楼相对，清末民初以来，社会上广泛流传着乾隆帝回妃登宝月楼与其族人隔街相望、排解乡愁的传说。

宝月楼外回回营清真寺宣礼楼 / 约1907年

Xuanli Lou of a Hui Mosque, Approx. 1907

《威弧获鹿图》卷 / 故宫博物院藏

The Painting of Weihu Huolu, The Palace Museum

《威弧获鹿图》卷描绘了在一次围猎中，一位回装女子向乾隆帝亲递箭矢，后者一箭中的的场景。画中女子会是传说中的“香妃”吗？

圆明园

Yuanming Yuan (Old Summer Palace)

圆明园是清帝经营时间最长、

建筑规模最大、内部装潢最精的皇家园林。

从雍正到咸丰，五朝皇帝一年中的大部分时间都是在圆明园中度过的。

他们在这里处理政务、生活起居，

是清帝在北京名副其实的第二政治中心。

圆明园全盛时曾有“圆明五园”之称，

包括圆明园、长春园、绮春园、熙春园、春熙院，

总占地面积达六千亩以上，建筑面积则超过了二十万平方米，拥有大小景群百余处。

1860年10月18日，在遭到英法联军劫掠后，圆明园被英军放火焚毁。

圆明园规月桥 / 19世纪末

Guiyue Bridge of Old Summer Palace, Late 19th Century

这是迄今为止发现的为数不多的圆明园木结构建筑被毁前的影像之一，十分珍贵。拍摄地点在圆明园廓然大公。照片中这座精美的廊桥名“规月桥”，是清帝乘舟出入双鹤斋的必经之地。

1860年10月18日英军火烧圆明园后，尚有北部部分建筑幸免于难。此帧照片反映了1860至1900年间，圆明园残存建筑的风貌。

圆明园廓然大公／1744年写景图

The Kuoran Dagong of Old Summer Palace, A Painting of Landscape in 1744

圆明园廓然大公在乾隆九年（1744 年）时尚无规月桥。此组建筑位于圆明园福海西北隅，因主殿名双鹤斋，清宫多以双鹤斋称呼这里。

长春园法慧寺多宝琉璃塔 / 19世纪末

Duobao Glazed Pagoda of Fahui Si in Changchun Yuan (Garden of Eternal Spring), Late 19th Century

此帧照片所示之多宝琉璃塔，在圆明园属园之一的长春园法慧寺内。其建筑形制与本篇后所附清漪园、静明园的两座多宝琉璃塔相似，但又有区别。法慧寺琉璃塔顶部为三重檐圆顶，塔身八方，寓意着“天圆地方”。而静明园、清漪园琉璃塔顶均为方顶，后两塔保存至今。法慧寺琉璃塔虽然早已无存，但其特殊的建筑形式仍然可以在紫禁城梵华楼内一座同时期建造的珐琅塔身上见到。

紫禁城梵华楼内珐琅塔 / 乾隆三十九年（1774年）造

Falang Pagoda of Fanhua Lou (The Buddhist Building), the 39th Year of Qianlong Emperor (1774)

长春园法慧寺多宝琉璃塔 / 19世纪末

Duobao Glazed Pagoda of Fahui Si in Changchun Yuan (Garden of Eternal Spring), Late 19th Century

清漪园（颐和园）

Qingyi Yuan (Summer Palace)

乾隆十五年（1750 年），乾隆帝为祝釐崇庆皇太后六旬万寿，
特命工部疏浚西湖，增饰瓮山，并系以“昆明湖”、“万寿山”的佳名。
在这片近 300 公顷的山水之间，倾十年之功，动用内帑上百万两，
成就了一座山水秀丽的皇家园林——清漪园。
清漪园的兴工与圆明园长春园类似，
从其最初的规划到园中每一处景致的细微雕琢，大到堆山造湖，
小到内檐装修，无一不体现着乾隆帝个人的园林品味。
与长春园一样，清漪园的每一个角落都被深深地烙上了移天缩地、踵事增华的印记。
乾隆十六年（1751 年），乾隆帝效法其祖康熙，
奉崇庆皇太后首度南巡。江南的自然与人文风情给他留下了深刻的印象。
一如康熙帝的南巡对畅春园工程的影响，
清漪园工程在乾隆帝南巡之后亦披上了一层浓重的江南风致。
至乾隆二十九年（1764 年），清漪园工程全部告竣之时，
园中明确写仿江南的园林景观即有佛香阁、大报恩延寿寺、五百罗汉堂、
惠山园、花神庙、苏州街、凤凰墩、西堤及六桥等二十余处，就连万寿山、
昆明湖的山水形状都几乎是杭州孤山与西湖的变体再现。
再加上园中大小建筑在装修手法上对江南风格的追求，可以毫不夸张地说，
清漪园结合了北方园林山水磅礴的大气与南方园林精巧富丽的隽美，
是南北园林完美结合的典范。

万寿山昆明湖 / 20世纪初

The Kunming Hu (Kunming Lake) and Wanshou Shan (Longevity Hill),Early 20th Century

颐和园排云殿前铜龙 / 约1907年

A Bronze Dragon in front of Paiyun Dian (Hall of Pushing Clouds), Approx. 1907

颐和园排云门后二宫门 / 约1907年

Ergong Men behind the Paiyun Men (Gate of Pushing Clouds), Approx. 1907

1860 年，清漪园被英法联军焚毁后，同治帝曾拟拆其废料用于圆明园重修，后因朝野的一致反对与内府工料银两的拮据而作罢。1884 年，略具山水形状，稍加整理即可恢复旧观的清漪园，在光绪帝“奉养东朝”的旨意下开始了修复工程，慈禧太后改园名为“颐和园”。1900 年八国联军进占颐和园时，对其多有破坏。两年后，清廷再次修复。两次修复仅恢复了万寿山的前山与东宫部分，后山大片的清漪园旧筑依然是一片断壁残垣。

从佛香阁俯瞰排云门、排云殿 / 1915~1920年

A Bird's-Eye View of Paiyun Men and Paiyun Dian from Fo'xiang Ge, 1915-1920

光绪十二年（1886 年），在大报恩延寿寺大雄宝殿旧址上建起一座重檐大殿，名排云殿。殿名取自晋代诗人郭璞“神仙排云出，但见金银台”句。

颐和园佛香阁／约1907年

Fo'xiang Ge (Buddhist Incense Pagoda) of Summer Palace, Approx. 1907

清漪园时期，在现在的佛香阁处曾仿照南京大报恩寺塔建有一座九级砖塔，该塔在乾隆二十三年（1758 年）接近完工时，因歪闪被拆除。两年后，建成了一座比原塔略矮的木结构佛香阁。1860 年 10 月 18 日，佛香阁被英军焚毁，光绪时重建。重建后的佛香阁高 41 米，八面三层，外重檐四层，一至三层出廊，八角攒尖顶。内供西天接引佛。

清漪园智慧海／1860年10月

Zhihui Hai (Temple of Sea of Wisdom) of Qingyi Yuan, October 1860

智慧海是一座周身琉璃装饰的无梁殿，建在万寿山巅，象征着佛法的无边智慧像大海一样宽广。此帧照片拍摄于 1860 年 10 月，英法联军劫掠清漪园期间。

颐和园宝云阁／约1907年

Baoyun Ge (Belvedere of Precious Clouds) of Summer Palace, Approx. 1907

宝云阁建于清漪园时期，是乾隆时代皇家园林中仅有的两座“铜殿”之一。另一座在避暑山庄珠源寺，名宗镜阁，与宝云阁形制相同。宝云阁建于乾隆二十年（1755年），高7.5米，重207吨，面阔三间，重檐歇山顶，正脊置佛塔。铜亭所有构件均仿木构，工序复杂，工艺精湛。阁内槛墙上铸有工匠名字，从铸匠、凿匠、拨蜡匠、镟匠、锉匠到木匠，共40人，实属难得。铜亭所有门窗菱花隔扇心均为内外二层，因可以随意拆卸，至20世纪初已有丢失，其中有法国人购得，20世纪90年代被归安到原处。

颐和园游廊 / 20世纪初

You Lang (The Long Corridor) of Summer Palace, Early 20th Century

万寿山前沿湖建有游廊，东起邀月门，西至石丈亭，全长728米，共273间。遍施彩绘，美轮美奂。照片所示为养云轩南侧游廊影像。此帧为着色照片。

清漪园昙花阁 / 1860年10月

Tanhua Ge (Belvedere of Broad-Leaved Epiphyllum) of Qingyi Yuan, October 1860

昙花阁建在万寿山南麓，重檐三层，流光溢彩，它的建筑平面是极为特殊的六角星形，其类型多变的建筑形制与纷繁复杂的装饰风格在清代宫廷建筑中是绝无仅有的。此帧照片拍摄于 1860 年 10 月，英法联军劫掠清漪园期间。

颐和园谐趣园／1915~1920年

Xiequ Yuan (Garden of Harmonious Delight) of Summer Palace, 1915-1920

谐趣园初建时曾名惠山园。乾隆帝南巡至无锡，倾心于惠山寄畅园的美景，回銮后命人在万寿山加以仿建。嘉庆时惠山园改名谐趣园。

颐和园谐趣园知鱼桥 / 1915~1920年

Zhiyu Qiao (Understanding Fish Bridge) of Xiequ Yuan, 1915-1920

谐趣园知鱼桥之名典出《庄子》中庄周与惠施“濠上之辩”的故事，清帝在此观鱼、知乐。北海濠濮涧亦有一座与其形制相似的石桥。

清漪园花承阁多宝琉璃塔 / 1860年10月

Duobao Glazed Pagoda of Huacheng Ge (Belvedere of Flower Plinth), October 1860

花承阁在清漪园后山，其西侧建有多宝琉璃塔，色彩斑斓，八面玲珑，该塔与前文长春园法慧寺琉璃塔的区别主要在塔顶。其又与静明园圣缘寺琉璃塔形制相仿。此帧照片拍摄于1860年10月，英法联军劫掠清漪园期间。

清漪园文昌阁 / 1860年10月

Wenchang Ge (Belvedere of Wenchang) of Qingyi Yuan, October 1860

文昌阁建在昆明湖东岸，崇台雉堞，阁作三层，十字交脊，它的平面布局是藏密曼陀罗在建筑上的立体再现。从照片中发现，文昌阁二层曾安设一架巨大的自鸣钟表，表盘指针指向6:30。此帧照片拍摄于1860年10月，英法联军劫掠清漪园期间。

颐和园铜牛／约1907年

A Bronze Ox of Summer Palace, Approx. 1907

乾隆帝在清漪园东岸立铜牛一座，是为镇水而设。据《汉书》记载，汉长安城南有昆明池，两岸曾立石人二，分别象征牛郎与织女。对比清漪园昆明湖，西岸设耕织图织染局，东岸立铜牛，乾隆帝在这里有意“隐寓《汉书》之意”，将长安昆明池与北京昆明湖联系在一起。

颐和园知春亭 / 约1907年

Zhichun Ting (Pavilion of Perceiving Spring) of Summer Palace, Approx. 1907

清漪园建成后，乾隆帝曾游览该园达千次之多，是除圆明园外，游幸次数最多的皇家园林。且每次来园必对园中景物歌咏诗章，几十年来积攒的清漪园诗作超过千篇。一般时候，乾隆帝从圆明园西南的藻园门出发，或乘舟、或策马，多数时他喜欢驾着御马由石道驰骋至清漪园，入东宫门后在园中各处游赏。以舟代步时，乘坐“如在天上”或“卧游书室”两艘御舟，泛棹于昆明湖上。过午晚膳后，便策马返回圆明园，四十年来从未留宿一日。在一次与内廷大臣一起游览该园时，赴京述职的两江总督尹继善对乾隆帝晨来暮反的习惯喻作“驰驿观山”，乾隆帝对此颇为欣赏，特在御制诗中大大记上了一笔。

颐和园十七孔桥 / 1915~1920年

Shiqikong Qiao (The Seventeen-Arch Bridge) of Summer Palace, 1915-1920

十七孔桥横卧昆明湖上，连接昆明湖东岸与南湖岛，是清代皇家园林中规模最大的石桥。十七孔桥的南北两排望柱上，雕刻着形态各异的狮子544只，其数量甚至超过了卢沟桥。正中桥洞的南北两面，镌刻着乾隆皇帝题写的对联。

颐和园十七孔桥 / 约1907年

Shiqikong Qiao of Summer Palace, Approx. 1907

颐和园玉带桥 / 约1907年

Yudai Qiao (Jade Belt Bridge) of Summer Palace, Approx. 1907

玉带桥建在颐和园西堤上，为模仿杭州西湖的“西堤六桥”之一。玉带桥下河道连接着清漪园与静明园两座皇家园林。乾隆时，御舟从高耸的桥洞中驶过，向西直通玉泉山。

从万寿山远眺治镜阁 / 1873年

Zhijing Ge, from the angle of Wanshou Shan, 1873

拍摄者站在万寿山上，向西南方拍摄下这帧照片。那时昆明湖湖水已经干涸，远处的治镜阁孤悬在湖床上，在它左侧的是著名的玉带桥。

治镜阁劫后残影 / 1873年

The Relic of Zhijing Ge, after the looting of Anglo-French Army, 1873

清漪园治镜阁／1873年

Zhijing Ge of Qingyi Yuan, 1873

治镜阁建在清漪园西堤西侧的湖面上，四面环水，内外双垣，其建筑布局是藏密曼陀罗的立体再现。由于其建于水中，躲过了1860年10月英军的焚毁。1884年重修清漪园时，岛上建筑被拆除，用于颐和园的修建。

治镜阁复原图／陆伟测绘

A Recovered Painting on Zhijing Ge, Measured and Drew by Lu Wei

修复后的清漪园北宫门／19世纪80年代

Beigong Men (North Palace Gate) of Qingyi Yuan, 1880s

1884年，清漪园修复工程启动，虽然照片中的北宫门一带得到了重新葺治，但万寿山北麓依旧断壁残垣。

清漪园毁后残迹／1878年

The Relic of Qingyi Yuan, 1878

1860年10月18日，包括清漪园在内的三山五园被3000余名英军骑兵焚毁。照片所示为园毁18年后，清漪园东宫、万寿山前山一带的残迹。拍摄者站在勤政殿后的山坡上，近景从左至右依次是玉澜堂、夕佳楼、宜芸馆，均已沦为废墟；昆明湖上依然种植着大片莲藕，只是岸边不见了水木自亲和乐寿堂；万寿山前山东西一线游廊全部化为灰烬；大报恩延寿寺也仅剩下空旷的高台；只有转轮藏、众香界与智慧海等少量建筑残存下来。

静明园

Jingming Yuan

静明园坐落在京西风景优美的玉泉山上，
是清代三山五园中营建时间较早的皇家园林，
与畅春园的兴建时间相近。
康熙十九年（1680 年），康熙帝在辽金以来的“玉泉”胜地建立行宫，
两年后落成，初名“澄心园”，后改名“静明园”。
乾隆十五年（1750 年），随着清漪园工程的展开，
对静明园的修缮与扩建也被逐步提上日程。
至乾隆十八年（1753 年），静明园始成“十六景”之胜，
包括：廓然大公、芙蓉晴照、玉泉趵突、竹炉山房、圣因综绘、绣壁诗态、
溪田课耕、清凉禅窟、采香云径、峡雪琴音、玉峰塔影、
风篁清听、镜影涵虚、裂帛湖光、云外钟声、翠云嘉荫。
乾隆二十四年（1759 年），园中主要工程告竣。
1860 年 10 月，在遭到英法联军劫掠后，静明园的部分建筑被英军焚毁。

静明园玉泉山南麓／19世纪末

The South Side of Yuquan Shan (Hill of Jade Fountain), Late 19th Century

这是一帧从静明园东宫门附近拍摄的玉泉山南麓照片。1860 年 10 月 18 日，英军火烧三山五园后，玉泉山南麓一些建筑并没有消失。照片中尚可辨认出定光塔、香岩寺、华严寺、香云法雨等建筑。

静明园玉峰塔影／20世纪初

A Pagoda on the Top of Yufeng (Jade Hill), Early 20th Century

此帧照片清晰地反映了静明园定光塔、香岩寺、华严寺 20 世纪初的风貌。照片右侧可见香云法雨一景北侧的华严洞。

静明园华藏海寺 / 1874年

Huazanghai Si (Huazanghai Temple) in Jingming Yuan, 1874

此帧照片拍摄于 1874 年，距静明园被焚毁仅 14 年。那时华藏海寺的木构建筑还有部分存留。寺内北侧建有七级八角密檐石塔。

静明园华藏海寺石塔 / 20世纪初

A Stone Pagoda in Huazanghai Si, Early 20th Century

静明园华藏海寺石塔 / 20世纪初

A Stone Pagoda in Huazanghai Si, Early 20th Century

静明园定光塔与华藏海寺石塔／20世纪初

Dingguang Pagoda and Stone Pagoda of Huazanghai Si, Early 20th Century

静明园峡雪琴音望妙高塔 / 约1907年

Miaogao Pagoda, from the angle of Xiaxue Qinyin, Approx. 1907

静明园“沁诗”花瓶门 / 20世纪初

“Qinshi” Huaping Men (Vase-shape Gate) in Jingming Yuan, Early 20th Century

花瓶门随墙而设，是皇家宫苑中的常见小品，在今天的北海公园、紫禁城乾隆花园里还能看到此类花瓶门。有意思的是，照片中门内一侧墙壁上，涂鸦着“胡世杰”等几字。熟悉宫廷历史的读者对此人当不陌生，他是乾隆皇帝身边的一名小太监，他的名字出现在一百多年后静明园的建筑上，恐怕是游园人的恶作剧，历史的巧合而已。

修复后的紫禁城延春阁花瓶门
林京摄于2014年

Huaping Men of Yanchun Ge in the Forbidden City, Photographed by Lin Jing, 2014

静明园玉辰宝殿 / 20世纪初

Yuchen Baodian (Yuchen Hall) of Jingming Yuan, Early 20th Century

静明园玉辰宝殿 / 20世纪初

Yuchen Baodian (Yuchen Hall) of Jingming Yuan, Early 20th Century

乾隆二十三年（1758年），玉泉山西麓建成一组大型宗教建筑群——仁育宫，又称东岳庙。玉辰宝殿在仁育宫内，砖石结构，俗称“无梁殿”。

静明园玉辰宝殿 / 20世纪初

Yuchen Baodian of Jingming Yuan, Early 20th Century

静明园圣缘寺多宝琉璃塔／20世纪初

Duobao Glazed Pagoda in Shengyuan Si of Jingming Yuan, Early 20th Century

圣缘寺在仁育宫南，寺后建有琉璃塔一座，形制与清漪园花承阁琉璃塔相仿。

静明园圣缘寺多宝琉璃塔／1915~1920年

Duobao Glazed Pagoda in Shengyuan Si of Jingming Yuan, 1915-1920

静明园清凉禅窟、仁育宫、圣缘寺远望／1900年

A Prospective View of Qingliang Chanku, Renyu Gong and Shengyuan Si, 1900

静宜园

Jingyi Yuan

静宜园地处京西大西山的香山东麓，
是三山五园中真正意义上占据真山真水的皇家园林之一。
乾隆八年（1743 年），乾隆帝在康熙帝建之香山寺行宫的基础上，
对香山东麓大加葺筑，将废弃多年的康熙行宫
扩展为一片占地面积达 140 余公顷的皇家园林，乾隆帝御赐园名“静宜园”。
静宜园工程始自乾隆八年（1743 年），
前期主要对“内垣”的建设为主。
两年后，工程深入到香山腹地，“外垣”工程渐次展开。
至乾隆十一年（1746 年）三月，静宜园成“二十八景”之胜。
即：勤政殿、丽瞩楼、绿云舫、虚朗斋、璎珞岩、翠微亭、青未了、
驯鹿坡、蟾蜍峰、栖云楼、知乐濠、香山寺、听法松、来青轩、
唳霜皋、香岩室、霞标磴、玉乳泉、雨香馆、绚秋林、晞阳阿、芙蓉坪、
香雾窟、栖月崖、重翠崦、玉华岫、森玉笏、隔云钟。
1860 年 10 月，在遭到英法联军劫掠后，静宜园被英军焚毁。

香山寺

勤政殿

静宜园勤政殿残迹 / 1922年

The Relic of Qinzheng Dian (Hall of Diligence to Politics), 1922

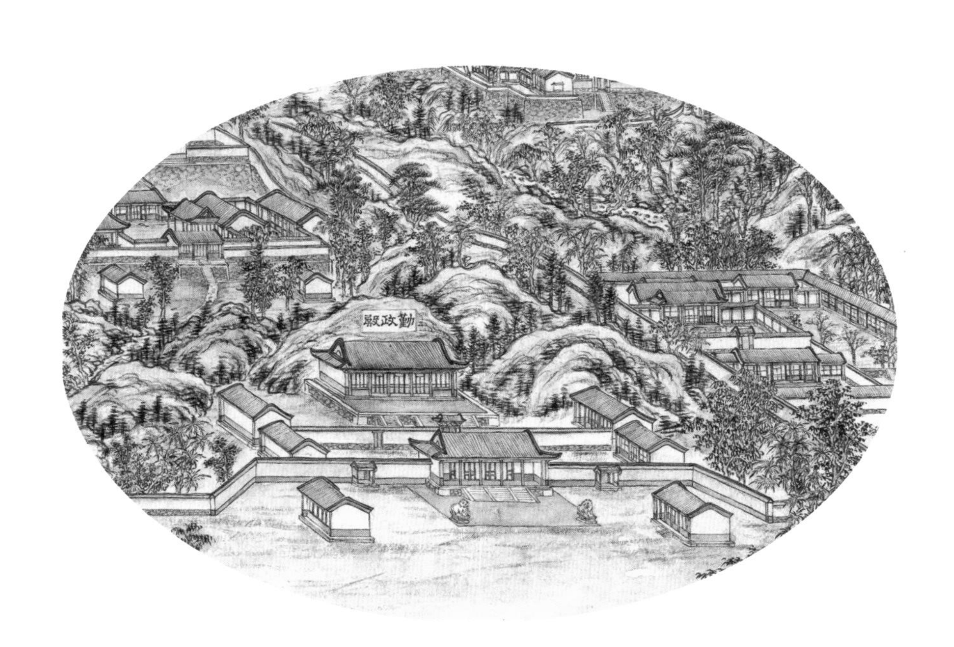

张若澄绘《静宜园二十八景图》卷 / 故宫博物院藏

The Painting on Twenty-Eight Scenes of Jingyi Yuan,
Drew by Zhang Ruocheng, The Palace Museum

静宜园宗镜大昭之庙前琉璃牌坊 / 1915~1920年

A Glazed Paifang in front of Zongjing Dazhao in Jingyi Yuan, 1915-1920

乾隆四十五年（1780年），乾隆帝为迎接六世班禅额尔德尼的到来，特在热河与北京兴建了两座仿照后藏扎什伦布寺的藏式庙宇作为班禅驻跸的行宫。位于北京静宜园中的一座名“宗镜大昭之庙”，六世班禅驻跸时，曾在此“开光念经”三日。

静宜园宗镜大昭之庙琉璃塔
1915~1920年
A Glazed Pagoda of Zongjing Dazhao in Jingyi Yuan, 1915-1920

静宜园宗镜大昭之庙琉璃塔／1915~1920年
A Glazed Pagoda of Zongjing Dazhao in Jingyi Yuan, 1915-1920

碧云寺五百罗汉堂内景／20世纪初

Interior Scene of Arhats Hall in Biyun Si (Temple of Azure Clouds), Early 20th Century

静宜园北有碧云寺，始建于元代，明清时期均有扩建。乾隆年间，乾隆帝命人在寺内仿海宁盐官安国寺建五百罗汉堂一座。在清代的皇家建筑中，曾经出现过三座五百罗汉堂，除了碧云寺外，清漪园、避暑山庄各建有一座，但保存至今的仅碧云寺一处。

碧云寺五百罗汉堂内景
1925~1949年

Interior Scene of Arhats Hall in Biyun Si, 1925-1949

碧云寺试泉悦性山房 / 20世纪初

A Tea Pavilion of Biyun Si,
Early 20th Century

碧云寺东侧建有一处环境幽致的院落，名“水泉院”。院中有乾隆帝引泉烹茶的凉榭——试泉悦性山房。

避暑山庄

The Summer Resort

避暑山庄是清帝在北京以外的又一处政治中心，
始建于康熙四十二年（1703年）。
从康熙到乾隆，避暑山庄大规模的营建时间在九十年以上，
共有大小景群百余处，较著名的是“康乾七十二景”。
在避暑山庄东、北两侧，还建有十二座规模不等的佛寺，
它们大多是清帝笼络少数民族与宗教上层人物的场所。
旧时，清帝每年在避暑山庄停留两月以上，期间还要赴木兰围场行围。
驻跸山庄时，内外蒙古、回部、西藏等少数民族首领，
以及朝鲜、安南、缅甸等属国使节均到此朝觐皇帝，
避暑山庄在历史上为维护多民族国家的统一与对外交往的繁荣做出了巨大的贡献。

避暑山庄丽正门／约1911年

Lizheng Men of the Summer Resort, Approx. 1911

丽正门是避暑山庄正门，乾隆三十六景之一。清帝驻跸山庄、木兰秋狝皆从此门出入。门前设影壁一座，左右立下马碑、挡众木。丽正门门额由乾隆帝用满、蒙、汉、藏、回五种文字书写，门后镌乾隆十九年（1754年）御制《丽正门》诗一首。

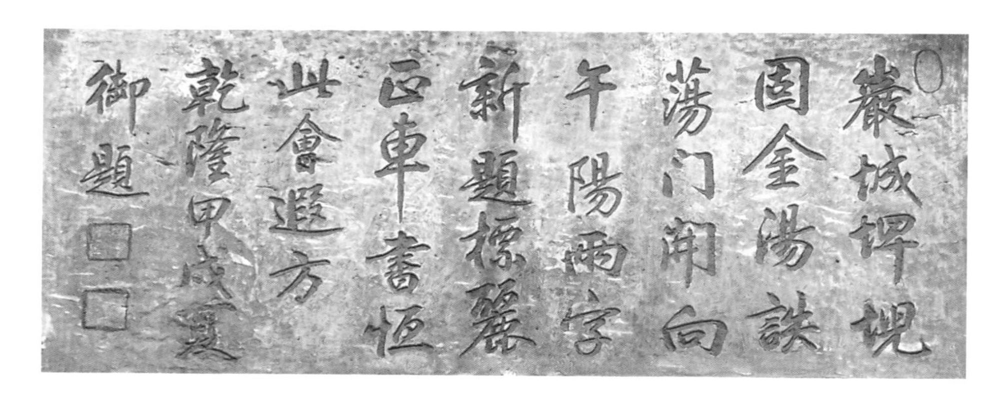

避暑山庄丽正门后乾隆帝御题《丽正门》诗

A Poem of Lizheng Men, behind the Lizheng Men of the Summer Resort, Wrote by Qianlong Emperor

避暑山庄万壑松风殿 / 约1911年

Wanhe Songfeng Dian (Hall of Wind through the Valley of Pines), Approx. 1911

万壑松风殿又名“纪恩堂”，康熙三十六景之一。据《热河志》记载，当年这里“长松数百，掩映周回”，而山庄千岩万壑，古松无数，塞风吹来，诚为“万壑松风”。这里曾是康熙帝引见官员、批阅章奏的地方。万壑松风殿也见证了康熙、乾隆二帝一段祖孙情深的历史往事。

康熙六十一年（1722 年），康熙帝携年仅 12 岁的弘历来到避暑山庄和木兰围场，这次出巡的经历给少年弘历留下了一段极为深刻的人生回忆，可以说影响了弘历的一生。

这年三月，在父亲胤禛的延请下，康熙帝来到圆明园，在牡丹台第一次见到了弘历。在众多孙辈中，只有弘历的俊俏和机敏博得了老皇帝的情有独钟，遂命人将其带至宫中亲自抚育。四月，銮驾起跸热河，这是老皇帝的第 55 次也是最后一次塞外之旅，特意将弘历带在身边。从四月到八月，在山庄的三个多月里，弘历被赐居万壑松风殿的“抑斋”读书。这里距离皇帝的寝宫不远，每当康熙有引见官员、批阅奏本之事，均让弘历陪侍一旁。据弘历事后回忆，祖父忙于公务时，自己便在一旁屏息凝视，亲身体会作为一国之君的风度与威严；在公务之余，或教授书写，或指导射箭，或频频赏赐御用膳食，祖孙俩甚至一同垂钓，将钓来的鱼送给父亲胤禛。帝王之家能有如此寻常百姓的欢乐，真的是难能可贵。

八月，康熙帝启程围场，开始了为期二十天的“木兰秋狝”。一天，皇帝在永安莽喀（后来乾隆帝命此地为“第一围场”，意为“沙岗”。）用火枪射倒一只熊，命侍卫带弘历去擒获，意在让他初次入围便得获熊的成绩。不料，受伤的熊突然发作扑向正欲上马的弘历，紧急关头康熙急忙用虎枪刺去，才算化解了一场危机。回到帐中，康熙深受触动，后怕之余指着弘历对温惠皇贵妃说：“他的命真是贵重啊！”此后不久，有猎人告说围内发现了一只老虎，弘历又抢着要去。鉴于前日的教训，康熙对他说：“你不能去，等到朕改日亲自入围时，再带你去吧。”爱护之情，溢于言表。

伴着万壑松风的陪侍，《爱莲说》的诵声，木兰围场的千钧一发和“福将过予”的祝愿，弘历陪伴着祖父度过了他人生的最后时光，在从避暑山庄回京的一个半月后，康熙帝在畅春园清溪书屋走到了人生的尽头。

这座曾经见证了祖孙一往情深的万壑松风殿，后来被弘历更名为“纪恩堂”，在他此后漫长的人生历程中，纪恩堂中往日的回忆未尝有过一丝一毫的消减，这在他后来大量的御制诗文中可以窥见一斑。

避暑山庄万树园 / 约1911年

Wanshu Yuan (Garden of Thousands of Trees) of the Summer Resort, Approx. 1911

避暑山庄万树园，乾隆三十六景之一。旧时，清帝在这里接见少数民族首领和外国使节，并在此举行大规模的蒙古帐宴。席间燃放“烟火盒子”，并安设转云游西洋秋千等娱乐表演设施。年节时还在此添设鳌山灯，用以点缀节日气氛。乾隆时，内附清廷的杜尔伯特部三策零、土尔扈特部渥巴锡、六世班禅额尔德尼以及第一个英国访华特使马戛尔尼等人均在这里得到了乾隆帝的接见。

避暑山庄万树园 / 约1911年

Wanshu Yuan of the Summer Resort, Approx. 1911

宽阔宏敞、树影婆娑的万树园和西侧的平冈试马埭，曾是清帝观看马术、校阅八旗子弟射箭的地方。乾隆十九年（1754年），乾隆帝在这里专门为归附朝廷的辉特部台吉阿睦尔撒那举行了一场声势浩大的马术表演。照片为冬日的万树园，远处可见白雪覆盖的烟雨楼和山庄湖区诸岛屿。

乾隆帝射箭挂屏 / 故宫博物院藏

An Oil Painging of Qianlong Emperor's Aiming a Target, The Palace Museum

这是一幅表现乾隆帝在避暑山庄试马埭练习步射的纪实油画。驻跸山庄时，乾隆帝在政务之余，会带领众皇子和八旗王公子弟到此，校阅他们的骑射、步射本领，并多会亲试弧矢，曾经有过发二十矢而中十九的佳绩。

避暑山庄鹫云寺／约1911年

Jiuyun Si of the Summer Resort, Approx. 1911

鹫云寺位于避暑山庄西峪深处，是一座小型庙宇。乾隆帝题额“鹫云寺”，意为灵鹫山云中的寺庙。灵鹫山是佛教圣地，将此寺比作灵鹫山的寺院，显示此寺的庄严、清净与神圣。照片中的三层六角建筑名“香界阁”，该寺现仅存遗址。

普宁寺全景 / 约1911年

A Panoramic Photograph of Puning Si (Temple of Universal Peace), Approx. 1911

普宁寺又称大佛寺，避暑山庄外八庙之一。乾隆二十年（1755 年），准噶尔部叛乱首领达瓦齐被生擒，缚献京师。这一年十月，乾隆帝在避暑山庄大宴厄鲁特四部贵族，并为纪念平准过程中的这一阶段性胜利，命人以西藏桑耶寺为蓝本在避暑山庄东北兴建了普宁寺。

普陀宗乘之庙全景 / 约1911年

A Panoramic Photograph of Putuo Zongcheng Miao (Temple of Potala), Approx. 1911

普陀宗乘之庙又称小布达拉宫，避暑山庄外八庙之一。为庆祝崇庆皇太后八旬万寿，乾隆帝命人在避暑山庄之北仿照拉萨布达拉宫兴建普陀宗乘之庙，工程历时四年，至乾隆三十六年（1771 年）竣工。当年，游牧在伏尔加河流域的土尔扈特部蒙古人众，在首领渥巴锡的带领下万里东归，乾隆帝在普陀宗乘之庙的主殿万法归一殿内接见了渥巴锡及众土尔扈特部上层，并在此举行了盛大的宗教法会。

乾隆帝普宁寺佛装像 / 故宫博物院藏

A Portiait of Qianlong Emperor Wearing Kasaya in Puning Si, The Palace Museum

须弥福寿之庙全景 / 1933年

A Panoramic Photograph of Xumi Fushou Miao (Temple of Sumeru Happiness and Longevity), 1933

须弥福寿之庙又称班禅行宫，避暑山庄外八庙之一。乾隆四十五年（1780 年）七月，六世班禅额尔德尼不远万里，从后藏扎什伦布寺启程，赴热河避暑山庄祝釐乾隆帝的七旬万寿。为迎接班禅大师，乾隆帝特在避暑山庄之北仿照扎什伦布寺兴建了须弥福寿之庙，作为班禅驻跸热河期间的行宫。照片右上角是普宁寺。

六世班禅坐像 / 故宫博物院藏

A Portiait of 6th Panchen Lama, The Palace Museum

农事试验场

Experimental Farm

光绪三十二年（1906 年），清政府于西直门外旧乐善园、
继园址和附近地方划地建农事试验场，效欧美日本，以求“振兴农务”。
至 1908 年，陆续建成了试验室、标本室、陈列室、照相馆、动物园等配套建筑。
其中，动物园于 1908 年对外开放。

农事试验场正门 / 1906年

The Front Gate of Experimental Farm, 1906

农事试验场正门东侧门楼 / 1906年

East Office House of Experimental Farm, 1906

农事试验场接待室 / 1906年

The Reception Building of Experimental Farm, 1906

农事试验场观稼轩与花圃 / 1906年

Guanjia Xuan and Flower Field in Experimental Farm, 1906

农事试验场动物园门 / 1906年

Gate of Zoo in Experimental Farm, 1906

畅观楼 / 1908年

Changguan Lou, 1908

畅观楼竣工于光绪三十四年（1908 年）年初，是农事试验场西侧一座造型别致的欧式洋楼。楼内铺设地毯，陈设大量西洋家具，四壁悬挂着螺钿、细绣挂屏等。楼成后，慈禧太后、光绪帝都曾登楼游赏。

黑龙潭龙王庙

The Dragon Temple of Heilong Spring

黑龙潭龙王庙位于海淀寿安山北麓，金山口北。
山下有潭，山上建庙，名“敕建黑龙王庙”。
乾隆帝敕封庙中龙王“昭灵沛泽龙王之神”。
旧时，每遇旱情，皇帝要至此向龙王祈雨，
待到龙王“显灵”而雨水沾足时，
又会回来向龙王“谢祈”，由此形成制度。

黑龙潭龙王庙远眺／20世纪初

A Prospective View of the Dragon Temple, Early 20th Century

黑龙潭龙王庙碑亭／1900年

Bei Ting (a pavilion built over a stone tablet) of the Dragon Temple, 1900

黑龙潭龙王庙花墙／1900年

The Flower Wall of the Dragon Temple, 1900

乐净山斋

Lejing Shanzhai

乐净山斋是逊帝溥仪的英文教师庄士敦
在妙峰山的私人别墅，建于北洋政府时期。
1919 年，庄士敦开始担任溥仪的英文教师，在宫中授课。
溥仪深受庄士敦的影响，
赏给他“头品顶戴”、“毓庆宫行走”、
“紫禁城内乘肩舆”等一连串的殊荣。
溥仪特为庄士敦的妙峰山别墅题写了“乐净山斋”匾额。

妙峰山庄士敦别墅 / 20世纪20年代

A Villa of Sir Reginald Fleming Johnston on Miaofeng Shan, 1920s

妙峰山庄士敦别墅 / 20世纪20年代

A Villa of Sir Reginald Fleming Johnston on Miaofeng Shan, 1920s

溥仪为庄士敦别墅题写的“乐净山斋”匾额 / 20世纪20年代

A Tablet with the Characters of Lejing Shanzhai in Sir Reginald Fleming Johnston's Villa, Wrote by Puyi, 1920s

醇亲王园寝

Tomb of Prince Chun

醇亲王园寝位于海淀阳台山，是道光帝第七子醇亲王奕譞的陵寝。
园寝选址妙高峰主峰下，这里是金章宗时期修建的“八大水院”之一的香水院旧址。
光绪十六年（1891年），醇亲王去世后，慈禧太后拨银五万两葺治园寝。
园寝依山势由东向西建有碑亭、
月牙河、石拱桥、园寝门、享殿、宝顶等建筑。
园寝内除了醇亲王奕譞和福晋叶赫那拉氏外，
还埋葬着三位侧室颜札氏、刘佳氏、李佳氏。
在醇亲王园寝西侧的山坡上，
还有醇亲王题写的“一卷永镇”、“漱石枕流”等山石题刻。
另在园寝东南侧山下，还建有一座规模较小的墓园，
埋葬着醇亲王早年夭折的几位儿女。

醇亲王园寝石阶／20世纪20年代

Stone Steps of the Tomb of Prince Chun, 1920s

醇亲王园寝碑亭 / 20世纪20年代

A Bei Ting of the Tomb of Prince Chun, 1920s

醇亲王奕譞是光绪皇帝的本生父，故其园寝碑亭使用了明黄色琉璃瓦覆顶，是清代亲王园寝仅见的一例。亭内石碑上镌刻着光绪帝御书碑文。

醇亲王园寝门 / 20世纪20年代

Gate of the Tomb of Prince Chun, 1920s

醇亲王园寝享殿 / 20世纪20年代

Xiang Dian (Xiang Hall) of the Tomb of Prince Chun, 1920s

醇亲王园寝主宝顶 / 20世纪20年代

The Grave Mound of the Tomb of Prince Chun, 1920s

主宝顶地宫内埋葬着醇亲王奕譞与福晋叶赫那拉氏，主宝顶由一圈雕刻精美的汉白玉须弥座承托。

醇亲王园寝阳宅“退潜别墅”／20世纪20年代

The Villa for Residence after Retirement of Prince Chun, 1920s

醇亲王奕譞将园寝北侧的阳宅建筑定名为“退潜别墅”，意在向朝廷表明自己虽然贵为光绪帝的本生父亲，但绝不会贪恋权位的与世无争的心态。整座阳宅建筑倚园寝北墙而建，前开城关门，院内五进，后设假山庭院，曲水回廊。院内厅房曾是奕譞生前来妙峰山过“退隐”生活的地方。奕譞去世后，他的后人每到园寝拜谒，也住在阳宅内。

醇亲王园寝阳宅“退潜别墅”院内／20世纪20年代

The Yard of the Villa for Residence after Retirement of Prince Chun, 1920s

此帧照片为“退潜别墅”的第二进院，右侧是别墅正殿纳云堂，堂前南北各有三间配殿。远景可见园寝北墙。

醇亲王园寝阳宅“退潜别墅”悦性亭 / 20世纪20年代

Yuexing Ting of the Villa for Residence after Retirement of Prince Chun, 1920s

“退潜别墅”内有一座造型别致的廊亭，亭中仿曲水流觞之意铺设一条流杯渠。这里是醇亲王过“退潜”生活的园林小品。

清西陵

West Mausoleum of Qing dynasty

清西陵位于河北易县永宁山下，东临拒马河，
南傍易水河，北面山峦起伏，灵岩岫翠，陵域广大，
是清代帝王在关内的第二座皇家陵园。
清西陵始建于雍正八年（1730 年），
最后一座陵寝完工于民国四年（1915 年），历时一百八十五年。
共建有帝陵四座（雍正帝泰陵、嘉庆帝昌陵、道光帝慕陵、光绪帝崇陵）、
皇后陵三座、妃园寝三座，另有亲王、阿哥、公主园寝四座。
西陵诸陵内，埋葬着四位皇帝、九位皇后、五十七位妃嫔、
六位亲王阿哥公主，共计七十六人。

清西陵泰陵石牌坊
约1907年
The Stone Paifang of Tai Ling of West Mausoleum of Qing dynasty, Approx. 1907

清西陵泰陵石牌坊
约1907年
The Stone Paifang of Tai Ling, Approx. 1907

作为清西陵首陵的泰陵，是雍正皇帝及其孝敬皇后、敦肃皇贵妃的陵寝，位于永宁山主峰之下。泰陵陵区规模宏大，体系完整，主要建筑包括：五孔桥、三路石牌坊、大红门、具服殿、圣德神功碑楼、七孔桥、神道、石像生、龙凤门、三孔桥、碑亭、隆恩门、隆恩殿、二柱门、石五供、方城明楼、宝顶。

清西陵泰陵龙凤门
约1907年
The Longfeng Men of Tai Ling, Approx. 1907

清西陵泰陵七孔桥和圣德神功碑楼
约1907年
The Stone Bridge and Shengde Shengong Beilou (Beilou of Great Merits) of Tai Ling, Approx. 1907

清西陵泰陵隆恩殿／约1907年

Long'en Dian of Tai Ling, Approx. 1907

清西陵昌陵华表 / 约1907年

A Huabiao of Chang Ling of West Mausoleum of Qing dynasty, Approx. 1907

清西陵昌陵石像生 / 约1907年

The Stone Statuaries of Chang Ling, Approx. 1907

昌陵建在太平峪，是清西陵中的第二座帝陵，埋葬着嘉庆皇帝和孝淑睿皇后。

清西陵昌陵全景／约1907年

A Panoramic Photograph of Chang Ling, Approx. 1907

清西陵慕陵三孔桥与碑亭 / 约1907年

The Stone Bridge and Bei Ting in Mu Ling of West Mausoleum of Qing dynasty, Approx. 1907

慕陵是道光皇帝与孝穆、孝慎、孝全三位皇后的陵寝。最初的道光皇帝陵曾选址在东陵绕斗峪，历七年建成。不料在葬入孝穆皇后之后，发现地宫渗水，经过修治仍不能解决问题，于是道光帝命人改在西陵陵域内重新选址建陵。道光十六年（1836年），西陵龙泉峪新陵完工，历时四年。在道光帝力主节俭的要求下，新陵的修建大大缩减了以往清代帝陵的规制，分别裁撤了圣德神功碑楼、华表、石像生、方城、明楼等建筑；隆恩殿改重檐为单檐；不设汉白玉栏板；鼎炉改设一对；裁撤铜鹿、铜鹤等。

道光帝虽然在陵寝规制上较他的祖先大为减省，但从东陵到西陵，两建一拆，所耗费的银两也是十分惊人的。

清西陵慕陵隆恩殿与石坊／约1907年
Long'en Dian and Stone Paifang of Mu Ling, Approx. 1907

清西陵慕陵石坊 / 约1907年

A Stone Paifang of Mu Ling, Approx. 1907

清西陵慕陵石五供与宝顶 / 约1907年

The Altar with Five Precious Objects and the Grave Mound of Mu Ling, Approx. 1907

崇陵工程纪实之
崇陵工作第一图 / 1909年
The Construction of Chong Ling (Part 1), 1909

照片中的白色柱状物，是工程开始之初，施工人员为标示陵寝主要建筑位置和实际高度而提前垒砌的“线墩”。

崇陵工程纪实之
崇陵工作第二图 / 1909年
The Construction of Chong Ling (Part 2), 1909

崇陵是光绪帝和隆裕皇后的陵寝，清西陵中最后一座皇陵。于宣统元年（1909 年）始建，其建筑布局、规制与同治帝惠陵相近。崇陵工程跨越清政府与北洋政府交替统治时期。1912 年 2 月，清帝退位后工程一度停止，后经国民政府与逊清内务府协商，以“皇室经费”继续赶修，主要工程延续了五年之久。1913 年 11 月 16 日，光绪帝、隆裕皇后梓宫奉安入崇陵地宫。这套工程纪实照片完整、清晰地再现了 1909 年至 1913 年间，崇陵工程从施工测量到竣工的全过程。同时，还包括对沿途御道、行宫、内务府、八旗、绿营营房等处新建与维修的照片实录。这批照片，为研究清代大型皇家工程各个阶段的施工程序与做法提供了第一手资料，显得弥足珍贵。

崇陵工程纪实之
崇陵工作第三图 / 1909年
The Construction of Chong Ling
(Part 3), 1909

崇陵工程纪实之
崇陵工作第四图 / 1909年
The Construction of Chong Ling
(Part 4), 1909

崇陵工程纪实之大桥
1909~1913年
The Construction of Chong Ling: Bridge, 1909-1913

崇陵工程纪实之
崇陵全图正面 / 1909~1913年
The Construction of Chong Ling: A Front View of Chong Ling, 1909-1913

崇陵工程纪实之
崇陵全图侧面／1909~1913年
The Construction of Chong Ling:
A Side View of Chong Ling,
1909-1913

崇陵工程纪实之石牌楼
1909~1913年
The Construction of Chong Ling:
Stone Pailou, 1909-1913

崇陵工程纪实之隆恩门
1909~1913年
The Construction of Chong Ling: Long'en Men, 1909-1913

崇陵工程纪实之隆恩殿
1909~1913年
The Construction of Chong Ling: Long'en Dian, 1909-1913

崇陵工程纪实之东配殿
1909~1913年
The Construction of Chong Ling: East Wing Hall, 1909-1913

崇陵工程纪实之琉璃门
1909~1913年
The Construction of Chong Ling: Liuli Men (Glazed Gate), 1909-1913

崇陵工程纪实之明楼正面
1909~1913年
The Construction of Chong Ling: A Front View of Ming Lou (Soul Tower), 1909-1913

崇陵工程纪实之明楼侧面
1909~1913年
The Construction of Chong Ling: A Side View of Ming Lou, 1909-1913

寶頂

崇陵工程纪实之宝顶
1909~1913年
The Construction of Chong Ling: Grave Mound, 1909-1913

崇陵工程纪实之金券门
1909~1913年
The Construction of Chong Ling: Jinquan Men, 1909-1913

崇陵工程纪实之石床正面
1909~1913年
The Construction of Chong Ling: A Front View of a Stone Bed, 1909-1913

崇陵工程纪实之妃陵全图
1909~1913年
The Construction of Chong Ling: A View of Fei Ling (Mausoleum of Concubines), 1909-1913

崇陵工程纪实之妃陵宝顶
1909~1913年
The Construction of Chong Ling: the Grave Mound of Fei Ling, 1909-1913

崇陵竣工纪实之五孔桥 / 1913年

The Finale of Chong Ling: Stone Bridge, 1913

崇陵竣工纪实之神厨库 / 1913年

The Finale of Chong Ling: Shenchu Ku (The Warehouse of Sacred Kitchen), 1913

崇陵竣工纪实之牌楼门、碑亭 / 1913年

The Finale of Chong Ling: Pailou and Bei Ting, 1913

崇陵竣工纪实之大碑亭 / 1913年

The Finale of Chong Ling: Big Bei Ting, 1913

崇陵竣工纪实之三孔三路汉白玉石桥 1913年

The Finale of Chong Ling: the Marble Bridge of Three Way, Three Archs, 1913

崇陵竣工纪实之隆恩门 / 1913年

The Finale of Chong Ling: Long'en Men, 1913

崇陵竣工纪实之隆恩殿 / 1913年

The Finale of Chong Ling: Long'en Dian, 1913

崇陵竣工纪实之琉璃门正面 / 1913年

The Finale of Chong Ling: A Front View of Liuli Men, 1913

崇陵竣工纪实之琉璃门背面 / 1913年

The Finale of Chong Ling: A Back View of Liuli Men, 1913

崇陵竣工纪实之石台五供 / 1913年

The Finale of Chong Ling: The Altar with Five Precious Objects, 1913

崇陵竣工纪实之方城明楼 / 1913年

The Finale of Chong Ling: Ming Lou of Fang Cheng (Square City), 1913

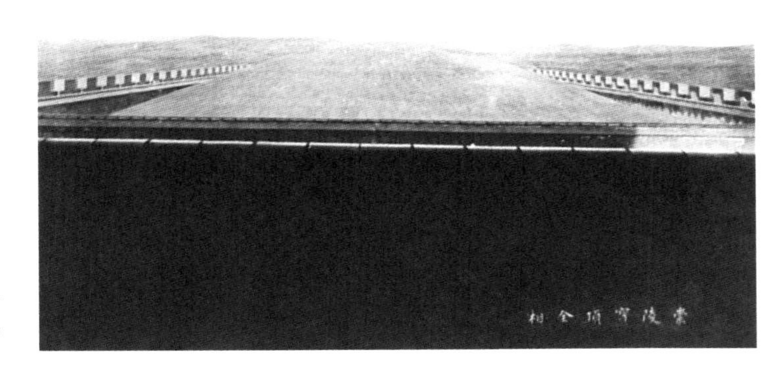

崇陵竣工纪实之宝顶 / 1913年

The Finale of Chong Ling: Grave Mound, 1913

崇陵竣工纪实之东沙子山 / 1913年

The Finale of Chong Ling: Dongshazi Shan (East Sand Mountain), 1913

崇陵竣工纪实之妃陵三孔桥 / 1913年

The Finale of Chong Ling: The Three-Arch White Stone Bridge of Fei Ling, 1913

拟修御路施工最要处第一（三岔口山西面，由南向北图）
1909~1911年

Program on Imperial Road Construction of West Mausoleum of Qing dynasty – 1: West Side of Sanchakou Hill, drawing from south to north, 1909-1911

宣统元年（1909 年）修建光绪帝陵时，沿途行宫、御道、内务府、八旗及绿营营房也在施工计划之中。以下为当时施工前的实测与施工中的纪实照片。

拟修御路施工最要处第二（三岔口山西面，由北向南图）
1909~1911年

Program on Imperial Road Construction of West Mausoleum of Qing dynasty – 2: West Side of Sanchakou Hill, drawing from north to south, 1909-1911

拟修御路施工最要处第三（三岔口山西面，由西南向东北旧有风水墙图）／1909~1911年

Program on Imperial Road Construction of West Mausoleum of Qing dynasty – 3: West Side of Sanchakou Hill, drawing from southwest to northeast of Fengshui Qiang (Geomantic Wall), 1909-1911

拟修御路施工最要处第四（三岔口山东面，由东北向西南图）
1909~1911年

Program on Imperial Road Construction of West Mausoleum of Qing dynasty – 4: West Side of Sanchakou Hill, drawing from northeast to southwest, 1909-1911

内务府大小圈营房地址 / 1909~1911年

The Location of the Barracks of Neiwu Fu (Imperial Household Department), 1909-1911

八旗大小圈营房地址 / 1909~1911年

The Location of the Barracks of Eight Banners, 1909-1911

绿营营房地址 / 1909~1911年

The Location of the Barracks of Green Camp, 1909-1911

修建中的梁各庄行宫宫门 / 1909~1911年

The Entrance Gate of the Imperial Palace of Liangge Zhuang, 1909-1911

梁各庄行宫是清帝展谒西陵沿途驻跸的四座行宫之一，位于西陵陵域以内。行宫建筑分东、中、西三路，旧有殿座上百间。1909 年，由于崇陵工程尚在施工中，光绪帝梓宫曾暂安梁各庄行宫直到 1913 年 11 月 16 日。照片为清末重修梁各庄行宫时留下的工程纪实照。

梁格庄
行宮大殿工作圖

修建中的梁各庄行宫大殿 / 1909~1911年

The Construction of the Main Hall of the Imperial Palace of Liangge Zhuang, 1909-1911

图版索引

上篇　从紫禁城到博物院

紫禁城外朝中路（含外三门）

大清门 / 26
大清门门额 / 27
中华门 / 28
中华门 / 28
中华门门额 / 29
天安门 / 30
天安门 / 31
天安门及门前华表 / 31
天安门前金水桥 / 32
天安门前石狮 / 32
天安门北侧 / 33
端门 / 34
午门 / 36
午门 / 38
午门 / 39
午门西雁翅楼 / 40
午门东雁翅楼 / 41
隆裕皇太后在建福宫花园 / 42
午门北侧（三联照）/ 43
太和门广场 / 45
太和门前石亭 / 46
太和门前青铜狮与熙和门 / 47
太和门前青铜狮 / 47
太和门 / 48
太和门 / 49
太和门内临时搭设的悼棚 / 49
太和门内景 / 49
太和殿广场 / 51
太和殿广场 / 52
太和殿广场 / 53
从太和门北侧望太和殿 / 54
太和殿 / 56
太和殿前丹陛石 / 57
太和殿前铜仙鹤 / 58
太和殿前铜龟 / 58
太和殿前嘉量 / 59
太和殿前日晷 / 59
太和殿内宝座与天花藻井 / 60
太和殿内景 / 61
太和殿内金漆宝座与雕龙髹金屏风 / 62
太和殿内袁世凯称帝时所用宝座 / 62
袁世凯所用宝座 / 62
太和殿内满铺的地毯 / 63
太和殿内紫宸台 / 64

太和殿内沥粉金柱 / 65
中和殿 / 66
中和殿内景 / 67
保和殿 / 68
保和殿内景 / 69
从乾清门看保和殿 / 70
保和殿后云龙大石雕 / 71

紫禁城外朝东路、西路
文华殿 / 74
文华殿记碑 / 74
文华殿 / 74
文渊阁 / 75
文渊阁 / 75
文渊阁二层 / 77
文渊阁明间内景 / 78
武英殿 / 80
武英殿 / 80
断虹桥 / 80
西华门北侧城墙修缮 / 81
南薰殿 / 81
西华门 / 82
《崇庆皇太后万寿庆典图》卷中的西华门及门外点景 / 82
东华门 / 83
角楼 / 84
角楼鸱吻 / 84
紫禁城西南角楼 / 85

紫禁城内廷中路
乾清门 / 88
乾清门 / 90
乾清门前镏金铜狮 / 91
乾清宫 / 92
乾清宫 / 93
乾清宫丹墀东侧 / 94
乾清宫前铜镏金江山社稷金殿 / 94
乾清宫前铜仙鹤 / 95
乾清宫前铜龟 / 96
乾清宫前铜香炉 / 97
乾清宫节日期间搭设的彩棚 / 98
乾清宫彩棚 / 98
乾清宫彩棚 / 99
乾清宫廊下安设的宫廷乐器 / 100
乾清宫廊下安设的宫廷乐器 / 100
乾清宫廊下安设的宫廷乐器 / 101
乾清宫内景 / 102
乾清宫内景 / 103
乾清宫内景 / 104
乾清宫宝座 / 105
乾清宫雕云龙纹镜 / 106
交泰殿 / 107
乾清宫东暖阁 / 107
节日时的交泰殿 / 108
节日时交泰殿张贴门神 / 109
交泰殿内蟠龙藻井 / 110
交泰殿内景 / 111
交泰殿内铜壶滴漏 / 112
交泰殿内大自鸣钟 / 112
圆明园慈云普护钟楼 / 113
坤宁宫 / 114
坤宁宫内萨满祭神处 / 115
坤宁宫节日时搭设的彩棚 / 116
坤宁宫东暖殿喜床 / 117
坤宁宫东暖殿明间内景 / 117
坤宁宫东暖殿东间内景 / 117
坤宁宫东暖殿东间内景 / 117
坤宁宫东暖殿东间陈设 / 118
坤宁宫东暖殿西间陈设 / 118
坤宁宫东暖殿毗卢帽上“日升月恒”匾额 / 119
坤宁宫东暖殿双开木板“喜”字门 / 120
庆宽绘《载湉大婚典礼全图》中的交泰殿与坤宁宫 / 121
天一门 / 122
天一门内连理柏 / 123
天一门前“海参”石 / 124

天一门前"诸葛拜斗"石 / 124
钦安殿 / 125
钦安殿抱厦内梁枋彩绘 / 126
钦安殿后檐墙与承光门 / 127
绛雪轩 / 128
绛雪轩立面图 / 129
绛雪轩南山墙 / 129
绛雪轩前琉璃花台 / 129
养性斋 / 130
养性斋内庄士敦书房 / 131
养性斋与四神祠 / 132
养性斋正立面图 / 133
养性斋前山石 / 134
养性斋前山石 / 134
养性斋前山石 / 135
千秋亭 / 136
千秋亭 / 136
万春亭 / 137
万春亭西侧 / 138
浮碧亭与摛藻堂 / 139
澄瑞亭东侧 / 139
御花园鹿囿 / 140
御花园古柏 / 141
御花园西井亭 / 142
延晖阁 / 143
延晖阁立面图 / 143
堆秀山 / 144
承光门 / 145
由堆秀山西望延晖阁 / 146
神武门 / 147
神武门 / 148

紫禁城内廷东路、西路

养心门 / 152
养心门外玉壁 / 154
养心门内 / 154
养心门外西值房 / 154
养心殿抱厦 / 154
养心殿抱厦 / 155
养心殿明间外景 / 155
养心殿明间内景 / 156
乾隆帝御制《新正养心殿》诗 / 156
养心殿明间蟠龙藻井 / 157
养心殿东暖阁内景 / 158
养心殿东暖阁内景 / 159
养心殿东暖阁内景 / 159
养心殿随安室 / 160
养心殿东暖阁内景 / 161
养心殿体顺堂内景 / 161
养心殿寝宫内景 / 162
养心殿东暖阁前檐炕床 / 162
养心殿院内花卉 / 163
养心殿院内花卉 / 163
养心殿院内花卉 / 163
斋宫抱厦内景 / 164
斋宫明间内景 / 164
斋宫殿内陈设 / 164
斋宫 / 165
毓庆宫东里间内景 / 165
承乾宫 / 166
永和宫抱厦 / 167
永和宫抱厦下的鹦鹉 / 167
永和宫后殿同顺斋内景 / 168
同顺斋婉容、文绣与溥仪姊弟们的合影 / 168
翊坤宫内景 / 169
翊坤宫前青铜露陈 / 169
体和殿东配殿寝床 / 170
储秀宫 / 171
储秀宫青铜露陈 / 172
储秀宫南窗炕床 / 172
储秀宫内浴缸 / 172
储秀宫 / 173
体元殿后抱厦 / 174
长春宫内景 / 174

西二长街嘉祉门 / 175
崇敬殿东间 / 175
重华宫东间陈设 / 176
重华宫东间陈设 / 176
漱芳斋外景 / 177

紫禁城内廷外东路
锡庆门 / 180
皇极殿 / 182
皇极殿西北侧 / 183
皇极殿 / 184
皇极殿内景 / 185
乾隆帝晚年朝服像 / 186
宁寿宫 / 187
养性门 / 188
养性门前鎏金铜狮 / 189
养性殿院落 / 190
乐寿堂 / 191
乐寿堂明间青玉“丹台春晓图”玉山 / 192
乐寿堂明间青玉云龙纹瓮 / 193
畅音阁戏楼 / 194
阅是楼内慈禧太后观戏处 / 195

紫禁城内廷外西路
慈宁宫 / 198
慈宁宫内景 / 198
《胪欢荟景图》册之《慈宁燕喜》 / 198
慈宁花园临溪亭 / 199
慈宁花园临溪亭 / 199
慈宁花园咸若馆 / 199
慈宁花园慈荫楼 / 199
慈宁花园 / 200
第一任故宫博物院院长易培基题临溪亭匾额 / 200
慈宁花园宝相楼 / 201
宝相楼内供奉的六座珐琅塔 / 201
春禧殿东配殿 / 202
春禧殿西配殿 / 202
寿安宫 / 203
《崇庆皇太后万寿庆典图》卷中的寿安宫三层大戏楼 / 203
寿安宫配楼 / 204
寿安宫西侧配楼 / 205
雨花阁四层佛龛供案 / 206
雨花阁 / 206
雨花阁 / 207
由宝华殿望雨花阁 / 208
梵宗楼 / 208
雨花阁北面 / 209
宝华殿前铜香炉 / 210
宝华殿内景 / 211
宝华殿法器 / 211
中正殿内供案 / 211
“扮鬼”的喇嘛 / 211
建福宫花园延春阁 / 212
由延春阁望北海琼华岛 / 213
建福宫花园积翠亭与广生楼 / 214
建福宫花园存性门火后残迹 / 214
建福宫花园延春阁火后残迹 / 215
建福宫花园虎皮墙圆光门火后残迹 / 215

下篇　皇家苑囿与陵寝
景山
景山与东三座门 / 226
景山万春亭 / 227
景山观妙亭 / 228
景山辑芳亭 / 228
景山辑芳亭 / 229
景山辑芳亭 / 229
寿皇门前西侧牌坊 / 230
寿皇门前东侧牌坊与石狮 / 232
寿皇殿 / 233
寿皇殿与丹陛石 / 233
寿皇殿内景 / 234
景山北望 / 235

西苑三海

由中海北望琼华岛 / 238
从积翠坊望琼华岛 / 239
金鳌玉蝀桥 / 240
北海团城承光殿内玉佛 / 242
北海善因殿 / 243
北海西天梵境前琉璃牌坊 / 244
北海九龙壁 / 246
从北海五龙亭望琼华岛 / 247
北海静心斋 / 247
中海紫光阁 / 248
乾隆帝《紫光阁赐宴图》 / 248
中海紫光阁 / 249
紫光阁明间内景 / 249
中海万善殿 / 250
中海水云榭 / 251
张若澄绘、乾隆御题《燕京八景图》册之太液秋风 / 251
中海海晏堂宫门 / 252
中海海晏堂 / 254
中海海晏堂明间内景 / 255
长春园海晏堂铜版画 / 255
南海瀛台 / 256
南海翔鸾阁 / 257
翔鸾阁叠落廊 / 257
南海日知阁 / 258
《日知荟说》书影 / 259
南海流水音 / 259
南海宝月楼 / 260
宝月楼外回回营清真寺宣礼楼 / 261
《威弧获鹿图》卷 / 261

圆明园

圆明园规月桥 / 264
圆明园廓然大公 / 265
长春园法慧寺多宝琉璃塔 / 266
紫禁城梵华楼内珐琅塔 / 266
长春园法慧寺多宝琉璃塔 / 267

清漪园（颐和园）

万寿山昆明湖 / 270
颐和园排云殿前铜龙 / 270
颐和园排云门后二宫门 / 271
从佛香阁俯瞰排云门、排云殿 / 272
颐和园佛香阁 / 273
清漪园智慧海 / 274
颐和园宝云阁 / 275
颐和园游廊 / 276
清漪园昙花阁 / 278
颐和园谐趣园 / 280
颐和园谐趣园知鱼桥 / 281
清漪园花承阁多宝琉璃塔 / 282
清漪园文昌阁 / 284
颐和园铜牛 / 286
颐和园知春亭 / 287
颐和园十七孔桥 / 288
颐和园十七孔桥 / 289
颐和园玉带桥 / 289
从万寿山远眺治镜阁 / 290
治镜阁劫后残影 / 290
清漪园治镜阁 / 291
治镜阁复原图 / 291
修复后的清漪园北宫门 / 292
清漪园毁后残迹 / 294

静明园

静明园玉泉山南麓 / 298
静明园玉峰塔影 / 299
静明园华藏海寺 / 300
静明园华藏海寺石塔 / 301
静明园华藏海寺石塔 / 301
静明园定光塔与华藏海寺石塔 / 302
静明园峡雪琴音望妙高塔 / 303
静明园“沁诗”花瓶门 / 304
修复后的紫禁城延春阁花瓶门 / 304
静明园玉辰宝殿 / 305

静明园玉辰宝殿／305
静明园玉辰宝殿／306
静明园圣缘寺多宝琉璃塔／307
静明园圣缘寺多宝琉璃塔／307
静明园清凉禅窟、仁育宫、圣缘寺远望／307

静宜园
静宜园勤政殿残迹／311
张若澄绘《静宜园二十八景图》卷／311
静宜园宗镜大昭之庙前琉璃牌坊／312
静宜园宗镜大昭之庙琉璃塔／313
静宜园宗镜大昭之庙琉璃塔／313
碧云寺五百罗汉堂内景／314
碧云寺五百罗汉堂内景／314
碧云寺试泉悦性山房／315

避暑山庄
避暑山庄丽正门／318
避暑山庄丽正门后乾隆帝御题《丽正门》诗／318
避暑山庄万壑松风殿／319
避暑山庄万树园／320
避暑山庄万树园／321
乾隆帝射箭挂屏／321
避暑山庄鹫云寺／322
普宁寺全景／323
普陀宗乘之庙全景／324
乾隆帝普宁寺佛装像／324
须弥福寿之庙全景／325
六世班禅坐像／325

农事试验场
农事试验场正门／328
农事试验场正门东侧门楼／329
农事试验场接待室／330
农事试验场观稼轩与花圃／331
农事试验场动物园门／332
畅观楼／333

黑龙潭龙王庙
黑龙潭龙王庙远眺／336
黑龙潭龙王庙碑亭／337
黑龙潭龙王庙花墙／337

乐净山斋
妙峰山庄士敦别墅／340
妙峰山庄士敦别墅／341
溥仪为庄士敦别墅题写的“乐净山斋”匾额／341

醇亲王园寝
醇亲王园寝石阶／344
醇亲王园寝碑亭／345
醇亲王园寝门／346
醇亲王园寝享殿／347
醇亲王园寝主宝顶／348
醇亲王园寝阳宅“退潜别墅”／349
醇亲王园寝阳宅“退潜别墅”院内／350
醇亲王园寝阳宅“退潜别墅”悦性亭／351

清西陵
清西陵泰陵石牌坊／354
清西陵泰陵石牌坊／354
清西陵泰陵龙凤门／355
清西陵泰陵七孔桥和圣德神功碑楼／355
清西陵泰陵隆恩殿／356
清西陵昌陵华表／357
清西陵昌陵石像生／357
清西陵昌陵全景／358
清西陵慕陵三孔桥与碑亭／359
清西陵慕陵隆恩殿与石坊／360
清西陵慕陵石坊／361
清西陵慕陵石五供与宝顶／361
崇陵工程纪实之崇陵工作第一图／362
崇陵工程纪实之崇陵工作第二图／362
崇陵工程纪实之崇陵工作第三图／363
崇陵工程纪实之崇陵工作第四图／363

崇陵工程纪实之大桥 / 364
崇陵工程纪实之崇陵全图正面 / 364
崇陵工程纪实之崇陵全图侧面 / 365
崇陵工程纪实之石牌楼 / 365
崇陵工程纪实之隆恩门 / 366
崇陵工程纪实之隆恩殿 / 366
崇陵工程纪实之东配殿 / 367
崇陵工程纪实之琉璃门 / 367
崇陵工程纪实之明楼正面 / 368
崇陵工程纪实之明楼侧面 / 368
崇陵工程纪实之宝顶 / 369
崇陵工程纪实之金券门 / 369
崇陵工程纪实之石床正面 / 370
崇陵工程纪实之妃陵全图 / 370
崇陵工程纪实之妃陵宝顶 / 371
崇陵竣工纪实之五孔桥 / 372
崇陵竣工纪实之神厨库 / 372
崇陵竣工纪实之牌楼门、碑亭 / 372
崇陵竣工纪实之大碑亭 / 373
崇陵竣工纪实之三孔三路汉白玉石桥 / 373
崇陵竣工纪实之隆恩门 / 373
崇陵竣工纪实之隆恩殿 / 374
崇陵竣工纪实之琉璃门正面 / 374
崇陵竣工纪实之琉璃门背面 / 374
崇陵竣工纪实之石台五供 / 374
崇陵竣工纪实之方城明楼 / 375
崇陵竣工纪实之宝顶 / 375
崇陵竣工纪实之东沙子山 / 375
崇陵竣工纪实之妃陵三孔桥 / 375
拟修御路施工最要处第一 / 376
拟修御路施工最要处第二 / 377
拟修御路施工最要处第三 / 378
拟修御路施工最要处第四 / 379
内务府大小圈营房地址 / 380
八旗大小圈营房地址 / 381
绿营营房地址 / 382
修建中的梁各庄行宫宫门 / 383
修建中的梁各庄行宫大殿 / 384

Index

From the Imperial Palace to the Palace Museum

The Middle Section of Outer Court of the Forbidden City

Daqing Men / 26

The Tablet of Daqing Men / 27

Gate of China / 28

Gate of China / 28

The Tablet of Gate of China / 29

Tiananmen / 30

Tiananmen / 31

Tiananmen and Huabiao / 31

Golden River Bridge in front of Tiananmen / 32

The Imperial Guardian Lions in front of Tiananmen / 32

North Side of Tiananmen / 33

Duan Men / 34

Wu Men / 36

Wu Men / 38

Wu Men / 39

West Yanchi Lou of Wu Men / 4o

East Yanchi Lou of Wu Men / 41

Empress Dowager Longyu at the Jianfu Gong Garden / 42

North Side of Wu Men / 43

Taihe Men Square / 45

A Stone Pavilion of Taihe Men / 46

A Bronze Lion and Gate of Xihe in front of Taihe Men / 47

A Bronze Lion in front of Taihe Men / 47

Taihe Men / 48

Taihe Men / 49

A Temporary Funeral inside the Taihe Men / 49

Interior Scene of Taihe Men / 49

Taihe Dian Square / 51

Taihe Dian Square / 52

Taihe Dian Square / 53

Taihe Dian, from the angle of north Taihe Men / 54

Taihe Dian / 56

Danbi Shi in front of Taihe Dian / 57

A Bronze Crane in front of Taihe Dian / 58

A Bronze Turtle in front of Taihe Dian / 58

A Jia Liang in front of Taihe Dian / 59

A Sundial in front of Taihe Dian / 59

Throne, Caisson and Ceiling of Taihe Dian / 60

Interior Scene of Taihe Dian / 61

Golden Lacquer Throne and Screen Carved with Dragon in Taihe Dian / 62

Throne in Taihe Dian / 62

Throne in Taihe Dian / 62
Taihe Dian Covered with Carpet / 63
A Dais in Taihe Dian / 64
The Golden Lacquer Pillar in Taihe Dian / 65
Zhonghe Dian / 66
Interior Scene of Zhonghe Dian / 67
Baohe Dian / 68
Interior Scene of Baohe Dian / 69
Baohe Dian, from the angle of Qianqing Men / 70
The Marble Stone Carved with Clouds and Dragons
at the Backyard of Baohe Dian / 71

The Eastern and Western Section of Outer Court of the Forbidden City
Stela of Wenhua Dian / 74
Wenhua Dian / 74
Wenhua Dian / 74
Wenyuan Ge / 75
Wenyuan Ge / 75
Second Floor of Wenyuan Ge / 77
Interior Scene of the Bright Room in Wenyuan Ge / 78
Wuying Dian / 80
Wuying Dian / 80
Duanhong Bridge / 80
The Repairing Progress of North Side of Xihua Men / 81
Nanxun Dian / 81
Xihua Men / 82
Xihua Men and the Exterior Scene of Xihua Men from the Painting
of Empress Dowager Chongqing's Birthday / 82
Donghua Men / 83
Corner Tower / 84
Chiwen from Corner Tower / 84
The South-West Corner Tower of the Forbidden City / 85

The Middle Section of Inner Court of the Forbidden City
Qianqing Men / 88
Qianqing Men / 90
A Golden Lacquer Bronze Lion in front of Qianqing Men / 91
Qianqing Gong / 92
Qianqing Gong / 93
East Side of Dan Chi of Qianqing Gong / 94
The Golden Lacquer Bronze Hall of Jiangshan Sheji
in front of Qianqing Gong / 94
A Bronze Crane in front of Qianqing Gong / 95
A Bronze Turtle in front of Qianqing Gong / 96
A Bronze Xiang Lu in front of Qianqing Gong / 97
A Festive Staging in Qianqing Gong / 98
Festive Decorations in Qianqing Gong / 98
Festive Decorations in Qianqing Gong / 99
The Imperial Instruments under the Colonnade
of Qianqing Gong / 100
The Imperial Instruments under the Colonnade
of Qianqing Gong / 101
Interior Scene of Qianqing Gong / 102
Interior Scene of Qianqing Gong / 103
Interior Scene of Qianqing Gong / 104
The Throne in Qianqing Gong / 105
A Grand Sandalwood Mirror in Qianqing Gong / 106
Jiaotai Dian / 107
The East Warm Chamber of Qianqing Gong / 107
Festive Decorations of Jiaotai Dian / 108
The Poster of Door God Placed on Jiaotai Dian / 109
The Caisson of Pan in Jiaotai Dian / 110
Interior Scene of Jiaotai Dian / 111
A Bronze Water Clock in Jiaotai Dian / 112
A Giant Striking Clock in Jiaotai Dian / 112
Clock Tower of Old Summer Palace from Ciyun Puhu / 113
Kunning Gong / 114
The Room for Shanmanist Worship of Kunning Gong / 115
A Festive Staging in Kunning Gong / 116
The Wedding Bed of East Warm Chamber in Kunning Gong / 117
Interior Scene of East Warm Chamber in Kunning Gong / 117

Interior Scene of East Room in East Warm Chamber of Kunning Gong / 117
Interior Scene of East Room in East Warm Chamber of Kunning Gong / 117
The Interior Design of East Room in East Warm Chamber of Kunning Gong / 118
The Interior Design of West Room in East Warm Chamber of Kunning Gong / 118
A Tablet with Risheng Yueheng in East Warm Hall of Kunning Gong / 119
The Door with the Character Xi in East Warm Hall of Kunning Gong / 120
Kunning Gong and Jiaotai Dian in Paining of the Imperial Wedding of Zaitian / 121
Tianyi Men / 122
A Cypress Tree Looks Like a Couple Holding Hands / 123
A Stone of Sea Cucumbers in front of Tianyi Men / 124
A Stone of Zhuge Baidou in front of Tianyi Men / 124
Qin'an Dian / 125
The Painting on the Timber Beam of Qin'an Dian / 126
The Back Eaves of Qin'an Dian and Chengguang Men / 127
Jiangxue Xuan / 128
An Architectural Drawing of Jiangxue Xuan / 129
South Gable of Jiangxue Xuan / 129
Glazed Platform in front of Jiangxue Xuan / 129
Yangxing Zhai / 130
Study of Sir Reginald Fleming Johnston in Yangxing Zhai / 131
Yangxing Zhai and Sishen Ci / 132
An Architectural Drawing of Yangxing Zhai / 133
Stone Hill of Yangxing Zhai / 134
Stone Hill of Yangxing Zhai / 134
Stone Hill of Yangxing Zhai / 135
Qianqiu Ting / 136
Qianqiu Ting / 136
Wanchun Ting / 137
West Side of Wanchun Ting / 138
Fubi Ting and Chizao Tang / 139
East side of Chengrui Ting / 139
The Deer Garden of Yuhua Yuan / 140
An Ancient Cypress Tree in Yuhua Yuan / 141
Xijing Ting of Yuhua Yuan / 142
Yanhui Ge / 143
An Architectural Drawing of Yanhui Ge / 143
Duixiu Shan / 144
Chengguang Men / 145
Yanhui Ge, from the angle of the west side of Duixiu Shan / 146
Shenwu Men / 147
Shenwu Men / 148

The Eastern and Western Section of Inner Court of the Forbidden City

Yangxin Men / 152
Bronze Screen Wall with Jade outside the Yangxin Men / 154
Inside the Yangxin Men / 154
The West Guardian Room outside the Yangxin Men / 154
The Covered Corridor of Yangxin Dian / 154
The Covered Corridor of Yangxin Dian / 155
Exterior Scene of the Bright Room in Yangxin Dian / 155
Interior Scene of the Bring Room in Yangxin Dian / 156
Poem on Xinzheng Yangxin Dian, Wrote by Qianlong Emperor / 156
The Caisson of Pan in Yangxin Dian / 157
Interior Scene of the East Warm Chamber in Yangxin Dian / 158
Interior Scene of the East Warm Chamber in Yangxin Dian / 159
Interior Scene of the East Warm Chamber in Yangxin Dian / 159
Suian Shi of Yangxin Dian / 160
Interior Scene of the East Warm Chamber in Yangxin Dian / 161
Interior Scene of Tishun Tang in Yangxin Dian / 161
Interior Scene of the Bedroom in Yangxin Dian / 162
A Kang at the East Warm Chamber of Yangxin Dian / 162
The Flowers at the Yard of Yangxin Dian / 163
The Flowers at the Yard of Yangxin Dian / 163
The Flowers at the Yard of Yangxin Dian / 163

Interior Scene under the Covered Corridor
of Zhai Gong / 164
Interior Scene of the Bright Room in Zhai Gong / 164
The Interior Design of Zhai Gong / 164
Zhai Gong / 165
Interior Scene of the East Room in Yuqing Gong / 165
Chengqian Gong / 166
The Covered Corridor of Yonghe Gong / 167
A Cockatoo under the Covered Corridor of Yonghe Gong / 167
Interior Scene of Tongshun Zhai / 168
A Photograph of Wanrong, Wenxiu and Siblings of Puyi / 168
Interior Scene of Yikun Gong / 169
Bronze Lu Chen in front of Yikun Gong / 169
A Bed in the East Wing Hall of Tihe Dian / 170
Chuxiu Gong / 171
Bronze Lu Chen of Chuxiu Gong / 172
A Kang at the South Window of Chuxiu Gong / 172
A Bathtub in Chuxiu Gong / 172
Chuxiu Gong / 173
The Covered Corridor behind the Tiyuan Dian / 174
Interior Scene of Changchun Gong / 174
Jiazhi Men at the Xi'er Long Street / 175
East Room of Chongjing Dian / 175
The Interior Design of the East Room in Chonghua Gong / 176
The Interior Design of the East Room in Chonghua Gong / 176
Exterior Scene of Shufang Zhai / 177

The Outer Eastern Section of Inner Court of the Forbidden City

Xiqing Men / 180
Huangji Dian / 182
Northwest side of Huangji Dian / 183
Huangji Dian / 184
Interior Scene of Huangji Dian / 185
Qianlong Emperor in Court Dress / 186
Ningshou Gong / 187
Yangxing Men / 188
A Golden Lacquer Bronze Lion in front of Yangxing Men / 189
Courtyard of Yangxing Dian / 190
Leshou Tang / 191
A Shoushan Stone Carved with a Painting of Dantai Chunxiao
in the Bright Room of Leshou Tang / 192
A Green Jade Fuhai Carved with Clouds and Dragons
in the Bright Room of Leshou Tang / 193
The Peking Opera Theatre of Changyin Ge / 194
A Luohan Bed for Empress Dowager Cixi to Watch
Peking Opera in Yueshi Lou / 195

The Outer Western Section of Inner Court of the Forbidden City

Cining Gong / 198
Interior Scene of Cining Gong / 198
The Painting of Cining Yanxi from a series of paintings
of Luhuan Huijing / 198
The Linxi Ting of Cining Garden / 199
The Linxi Ting of Cining Garden / 199
Xianruo Guan of Cining Garden / 199
Ciyin Lou of Cining Garden / 199
Cining Garden / 200
Tablet of Linxi Ting, wrote by the first director
of the Palace Museum Yi Peiji / 200
The Baoxiang Lou of Cining Garden / 201
Six Porcelain Enamel Pagodas / 201
The East Wing Hall of Chunxi Dian / 202
The West Wing Hall of Chunxi Dian / 202
Shou'an Gong / 203
The Three-Floor Opera Theatre of Shou'an Gong from The
Painting of Empress Dowager Chongqing's Birthday / 203
Wing Building of Shou'an Gong / 204
West Wing Building of Shou'an Gong / 205
Four Tiers Shrine of Buddha in Yuhua Ge / 206
Yuhua Ge / 206

Yuhua Ge / 207
Yuhua Ge, from the angle of Baohua Dian / 208
Fanzong Lou / 208
North Side of Yuhua Ge / 209
A Xiang Lu in front of Baohua Dian / 210
Interior Scene of Baohua Dian / 211
The Ritual Implements of Baohua Dian / 211
The Altar in Zhongzheng Dian / 211
Lamas with Ghost Masks / 211
Yanchun Ge of Jianfu Gong Garden / 212
Qionghua Dao of Beihai, from the angle of Yanchun Ge / 213
Jicui Ting and Guangsheng Lou of Jianfu Gong Garden / 214
The Relic of Cunxing Men, after the fire
of Jianfu Gong Garden / 214
The Relic of Yanchun Ge, after the fire
of Jianfu Gong Garden / 215
The Relic of Hupi Wall and Yuanguang Men, after the fire
of Jianfu Gong Garden / 215

Imperial Gardens and Mausoleums

Jingshan

Jingshan and the East Three Gates / 226
Wanchun Ting of Jingshan / 227
Guanmiao Ting of Jingshan / 228
Jifang Ting of Jingshan / 228
Jifang Ting of Jingshan / 229
Jifang Ting of Jingshan / 229
Paifang at the West Side of Shouhuang Men / 230
Paifang and Stone Lions at the East Side
of Shouhuang Men / 232
Shouhuang Dian / 233
Shouhuang Dian and Danbi Shi / 233
Interior Scene of Shouhuang Dian / 234
The View from the North Side of Jingshan / 235

Xiyuan Sanhai (Sea Palace)

Qionghua Dao, from the angle of Zhonghai / 238
Qionghua Dao, from the angle of Jicui Fang / 239
Jin'ao Yudong Qiao / 240
Jade Buddha of Chengguang Dian in Tuan Cheng / 242
Shanyin Dian of Beihai / 243
A Glazed Paifang in front of Xitian Fanjing / 244
The Nine-Dragon Wall of Beihai / 246
Qionghua Dao, from the angle of Wulong Ting / 247
Jingxin Zhai / 247
Ziguang Ge / 248
A Dinner Granted by Qianlong Emperor at Ziguang Ge / 248
Ziguang Ge / 249
Interior Scene of the Bright Room in Ziguang Ge / 249
Wanshan Dian of Zhonghai / 250
Shuiyun Xie of Zhonghai / 251
Taiye Qiufeng from the Poetry "Eigth Scenes of Peking" / 251
The Entering Gate of Haiyan Tang / 252
Haiyan Tang of Zhonghai / 254
Interior Scene of the Bright Room in Haiyan Tang / 255
An Intaglio Painting on Haiyan Tang of Changchun Yuan / 255
Ying Tai of Nan Hai / 256
Xiangluan Ge of Nan Hai / 257
Dieluo Lang of Xiangluan Ge / 257
Rizhi Ge of Nanhai / 258
Liushui Yin of Nanhai / 259
A Photo of Rizhi Huishuo / 259
Baoyue Lou of Nanhai / 260
Xuanli Lou of a Hui Mosque / 261
The Painting of Weihu Huolu / 261

Yuanming Yuan (Old Summer Palace)

Guiyue Bridge of Old Summer Palace / 264
The Kuoran Dagong of Old Summer Palace / 265
Duobao Glazed Pagoda of Fahui Si in Changchun Yuan / 266

Falang Pagoda of Fanhua Lou / 266
Duobao Glazed Pagoda of Fahui Si in Changchun Yuan / 267

Qingyi Yuan (Summer Palace)
The Kunming Hu and Wanshou Shan / 270
A Bronze Dragon in front of Paiyun Dian / 270
Ergong Men behind the Paiyun Men / 271
A Bird's-Eye View of Paiyun Men and Paiyun Dian
from Fo'xiang Ge / 272
Fo'xiang Ge of Summer Palace / 273
Zhihui Hai of Qingyi Yuan / 274
Baoyun Ge of Summer Palace / 275
You Lang of Summer Palace / 276
Tanhua Ge of Qingyi Yuan / 278
Xiequ Yuan of Summer Palace / 280
Zhiyu Qiao of Xiequ Yuan / 281
Duobao Glazed Pagoda of Huacheng Ge / 282
Wenchang Ge of Qingyi Yuan / 284
A Bronze Ox of Summer Palace / 286
Zhichun Ting of Summer Palace / 287
Shiqikong Qiao of Summer Palace / 288
Shiqikong Qiao of Summer Palace / 289
Yudai Qiao of Summer Palace / 289
Zhijing Ge, from the angle of Wanshou Shan / 290
The Relic of Zhijing Ge, after the looting
of Anglo-French Army / 290
Zhijing Ge of Qingyi Yuan / 291
A Recovered Painting on Zhijing Ge / 291
Beigong Men of Qingyi Yuan / 292
The Relic of Qingyi Yuan / 294

Jingming Yuan
The South Side of Yuquan Shan / 298
A Pagoda on the Top of Yufeng / 299
Huazanghai Si in Jingming Yuan / 300
A Stone Pagoda in Huazanghai Si / 301
A Stone Pagoda in Huazanghai Si / 301
Dingguang Pagoda and Stone Pagoda of Huazanghai Si / 302
Miaogao Pagoda, from the angle of Xiaxue Qinyin / 303
"Qinshi" Huaping Men in Jingming Yuan / 304
Huaping Men of Yanchun Ge in the Forbidden City / 304
Yuchen Baodian of Jingming Yuan / 305
Yuchen Baodian of Jingming Yuan / 305
Yuchen Baodian of Jingming Yuan / 306
Duobao Glazed Pagoda in Shengyuan Si of Jingming Yuan / 307
Duobao Glazed Pagoda in Shengyuan Si of Jingming Yuan / 307
A Prospective View of Qingliang Chanku, Renyu Gong
and Shengyuan Si / 307

Jingyi Yuan
The Relic of Qinzheng Dian / 311
The Painting on Twenty-Eight Scenes of Jingyi Yuan,
Drew by Zhang Ruocheng / 311
A Glazed Paifang in front of Zongjing Dazhao
in Jingyi Yuan / 312
A Glazed Pagoda of Zongjing Dazhao in Jingyi Yuan / 313
A Glazed Pagoda of Zongjing Dazhao in Jingyi Yuan / 313
Interior Scene of Arhats Hall in Biyun Si / 314
Interior Scene of Arhats Hall in Biyun Si / 314
A Tea Pavilion of Biyun Si / 315

The Summer Resort
Lizheng Men of the Summer Resort / 318
A Poem of Lizheng Men, behind the Lizheng Men
of the Summer Resort, Wrote by Qianlong Emperor / 318
Wanhe Songfeng Dian / 319
Wanshu Yuan of the Summer Resort / 320
Wanshu Yuan of the Summer Resort / 321
An Oil Painging of Qianlong Emperor's Aiming a Target / 321
Jiuyun Si of the Summer Resort / 322
A Panoramic Photograph of Puning Si / 323
A Panoramic Photograph of Putuo Zongcheng Miao / 324

A Portiait of Qianlong Emperor Wearing Kasaya in Puning Si / 324
A Panoramic Photograph of Xumi Fushou Miao / 325
A Portiait of 6th Panchen Lama / 325

Experimental Farm
The Front Gate of Experimental Farm / 328
East Office House of Experimental Farm / 329
The Reception Building of Experimental Farm / 330
Guanjia Xuan and Flower Field in Experimental Farm / 331
Gate of Zoo in Experimental Farm / 332
Changguan Lou / 333

The Dragon Temple of Heilong Spring
A Prospective View of the Dragon Temple / 336
Bei Ting of the Dragon Temple / 337
The Flower Wall of the Dragon Temple / 337

Lejing Shanzhai
A Villa of Sir Reginald Fleming Johnston on Miaofeng Shan / 340
A Villa of Sir Reginald Fleming Johnston on Miaofeng Shan / 341
A Tablet with the Characters of Lejing Shanzhai in Sir Reginald Fleming Johnston's Villa, Wrote by Puyi / 341

Tomb of Prince Chun
Stone Steps of the Tomb of Prince Chun / 344
A Bei Ting of the Tomb of Prince Chun / 345
Gate of the Tomb of Prince Chun / 346
Xiang Dian of the Tomb of Prince Chun / 347
The Grave Mound of the Tomb of Prince Chun / 348
The Villa for Residence after Retirement of Prince Chun / 349
The Yard of the Villa for Residence after Retirement of Prince Chun / 350
Yuexing Ting of the Villa for Residence after Retirement of Prince Chun / 351

West Mausoleum of Qing dynasty
The Stone Paifang of Tai Ling of West Mausoleum of Qing dynasty / 354
The Stone Paifang of Tai Ling / 354
The Longfeng Men of Tai Ling / 355
The Stone Bridge and Shengde Shengong Beilou of Tai Ling / 355
Long'en Dian of Tai Ling / 356
A Huabiao of Chang Ling of West Mausoleum of Qing dynasty / 357
The Stone Statuaries of Chang Ling / 357
A Panoramic Photograph of Chang Ling / 358
The Stone Bridge and Bei Ting in Mu Ling of West Mausoleum of Qing dynasty / 359
Long'en Dian and Stone Paifang of Mu Ling / 360
A Stone Paifang of Mu Ling / 361
The Altar with Five Precious Objects and the Grave Mound of Mu Ling / 361
The Construction of Chong Ling (Part 1) / 362
The Construction of Chong Ling (Part 2) / 362
The Construction of Chong Ling (Part 3) / 363
The Construction of Chong Ling (Part 4) / 363
The Construction of Chong Ling: Bridge / 364
The Construction of Chong Ling: A Front View of Chong Ling / 364
The Construction of Chong Ling: A Side View of Chong Ling / 365
The Construction of Chong Ling: Stone Pailou / 365
The Construction of Chong Ling: Long'en Men / 366
The Construction of Chong Ling: Long'en Dian / 366
The Construction of Chong Ling: East Wing Hall / 367
The Construction of Chong Ling: Liuli Men / 367
The Construction of Chong Ling: A Front View of Ming Lou / 368
The Construction of Chong Ling: A Side View of Ming Lou / 368

The Construction of Chong Ling: Grave Mound / 369
The Construction of Chong Ling: Jinquan Men / 369
The Construction of Chong Ling: A Front View of a Stone Bed / 370
The Construction of Chong Ling: A View of Fei Ling / 370
The Construction of Chong Ling: the Grave Mound of Fei Ling / 371
The Finale of Chong Ling: Stone Bridge / 372
The Finale of Chong Ling: Shenchu Ku / 372
The Finale of Chong Ling: Pailou and Bei Ting / 372
The Finale of Chong Ling: Big Bei Ting / 373
The Finale of Chong Ling: the Marble Bridge of Three Way, Three Archs / 373
The Finale of Chong Ling: Long'en Men / 373
The Finale of Chong Ling: Long'en Dian / 374
The Finale of Chong Ling: A Front View of Liuli Men / 374
The Finale of Chong Ling: A Back View of Liuli Men / 374
The Finale of Chong Ling: The Altar with Five Precious Objects / 374
The Finale of Chong Ling: Ming Lou of Fang Cheng / 375
The Finale of Chong Ling: Grave Mound / 375
The Finale of Chong Ling: Dongshazi Shan / 375
The Finale of Chong Ling: The Three-Arch White Stone Bridge of Fei Ling / 375
Program on Imperial Road Construction of West Mausoleum of Qing dynasty – 1: West Side of Sanchakou Hill, drawing from south to north / 376
Program on Imperial Road Construction of West Mausoleum of Qing dynasty – 2: West Side of Sanchakou Hill, drawing from north to south / 377
Program on Imperial Road Construction of West Mausoleum of Qing dynasty – 3: West Side of Sanchakou Hill, drawing from southwest to northeast of Fenshui Qiang / 378
Program on Imperial Road Construction of West Mausoleum of Qing dynasty – 4: West Side of Sanchakou Hill, drawing from northeast to southwest / 379
The Location of the Barracks of Neiwu Fu / 380
The Location of the Barracks of Eight Banners / 381
The Location of the Barracks of Green Camp / 382
The Entrance Gate of the Imperial Palace of Liangge Zhuang / 383
The Construction of the Main Hall of the Imperial Palace of Liangge Zhuang / 384

编后记

近年来，伴随着传统文化在国内的强势回归，中国已经进入了一个全民收藏的年代。以往收藏领域的冷门——历史照片，也借助于时代，在一波波拍卖、一场场展览和市场上间或出现的动辄几万、十几万一张“原版照片”的新闻爆料中，正在以一种近乎狂热的状态不断地牵动人们的视线。这其中自然有公众对历史照片价值认知的日益提高，乐见于老照片传承古往、服务当代的价值属性得到不断彰显；又因买卖双方不对等的利益期待而不可避免地滤上一层商业炒作的色彩。

其实照片，尤其是历史照片，其价值就在摄影师有心抑或无意的快门瞬间留下的那个基于一张平面的时代，时间越久远，变迁越频繁，这张平面对今人而言就越有意义。就像我们每个人翻开相册，看到曾经青涩的自己，一个久违的声音，一缕远去的味道，都可以让我们重拾记忆，感受岁月的变迁，进而思考我们这一路从何走来？在照片的时间节点上，好像前人为我们打开了一扇时间之窗，如此洁净透明，百年光景，触手便可感知。通过这扇时间之窗，我们可以发出这样的豪迈：一张照片可以再现圆明园！这就是记录的意义所在。

所以，对每一张历史照片的信息尽可能全面的解读，使之服务当代、教育公众，较之拍卖会上虚妄浮夸的价格哄抬更具有现实意义。当然，我们还是希望中国的老照片市场能够健康有序的发展，继续为公众挖掘出更多、更精彩的尘封的历史记忆。换一个角度来说，以政治经济学的视角考察，任何时候都是价值来决定价格，亘古不变的道理。

《故宫藏影》的编辑出版，是故宫博物院和故宫出版社（原紫禁城出版社）在近三十年不断研究、公开旧藏历史照片的传统上出版的又一部规模宏大、内容新颖的重量级照片册。回顾故宫出版社相关题材的出版物，从 20 世纪 80 年代的专题画册，到 1990 年《故宫旧藏人物照片》、1995 年《帝京旧影》、1995 年《故宫珍藏人物照片荟萃》以及种类繁多的老照片系列明信片，可以说故宫博物院和故宫出版社是掀起 20 世纪八九十年代一波又一波老照片热的“策源地”。每一次院藏历史照片的公开披露，都会引起学术和公众领域不同程度的反响。

《故宫藏影》的最大特点，是在延续紫禁城题材的同时，跳出故宫的一方城池，将视野延展到皇家建筑、宫廷人物、洋务实业三个主题鲜明且内容新颖的领域。主题看似分散，内容却是高度统一：全部照片均围绕在清末至民国时期的皇室核心，建筑、人物、军事、经济均有触及。

《故宫藏影》涉及的历史照片，相当一部分来自故宫博物院图书馆的收藏。成书过程中，得到了张荣馆长、朱赛虹前馆长两位馆长的大力支持，图书馆曹莉、李英、武安国、刁美林等同仁在照片的整理工作中付出了心血，在他们的精心保管和严格筛选下，《故宫藏影》有了可靠的资源保障；在故宫博物院资深历史照片专家林京、左远波两位编审的细致考证下，《故宫藏影》有了过硬的学术保障；对本书的出版予以顾问支持的还有故宫出版社冯印淙、圆明园管理处刘阳、承德市文物局陈东、北京建工建筑设计研究院熊炜以及韩立恒、陆伟六位同志，在此一并致谢。

九十年的博物院，六百年的紫禁城；后之视今，亦犹今之视昔。希望我们这些终会被印在历史照片里的人，都能在各自的舞台上为后人留下我们这个时代的精彩。

编者

二〇一四年八月

Prologue

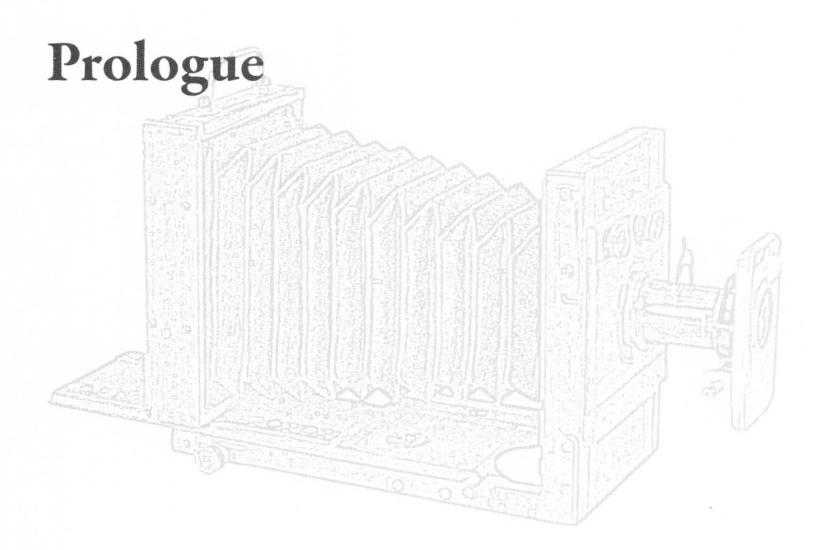

In recent years, China has started a renaissance of Chinese traditional culture. Later, the whole Chinese society has moved to a phase of collecting. Due to the popularity of Chinese traditional cultural value, the field of old photographs has run into the spotlights of people's daily life. Consequently, the headlines feature with the story of a ten thousand dollars' old photo have attracted huge attentions from society. This dramatic phenomena reflects two prospective of views, one is that the traditional cultural value is more acceptable to the Chinese society nowadays; the other is that the inequality of the commercial operation on photographic collection.

Talking about photographs, especially historical photographs, they could be defined as the two-dimensional history that captured by photographers; which means that the more the places that captured in a photograph have changed, the more meaningful the picture would be. When we open an old photo album, the pictures remind us all these nostalgic stories of our life. The photos make us to think of a question: where are we really from? Would these historical photographs show the roots of us? In fact, these old photographs offer us an opportunity to understand and learn the stories that happened 100 years ago. Furthermore, from the old photos, we even could say that the glamorous of Yuanming Yuan (the Old Summer Palace) will be revived from the old photos. That is the meaning of photographic recording.

Rather than the auction of photography, understanding and learning the information of every old photograph have further meanings to society. At the same time, we also hope that the market of old photographs has a healthier environment in terms of its development, and hope that the market could continue to discover the value of old pictures in order to unfold the history that we do not know. At the same Besides, in terms of the study of political economy, it always is that the value decides the price.

The publishing of *The Photographic Collection of the Palace Museum*, which edited by the Forbidden City Publishing House, based on the 30-year's research and study on old photographs. The book comprises the most creative content and the most unique collection of photographs. In terms of the photographic publishing of the Forbidden City Publishing House, the Forbidden City and publishing house are the birthplace of the popularity of the old photographs, drawing the timeline from *The Specialized Album of Paintings* on 1980s to the *Portraits of the Forbidden City* on 1990s; *The Old Photographs of the Imperial City* on 1995 and *The Classic Portraits from the Forbidden City* on 1995. Every time there is a new photographic publishing issued by the publishing house, it always receives a feedback from both academic and public fields.

Furthermore, one of the most significant features of *The Photographic Collection of the Palace Museum*, it is the review of the content. The book provides a new prospect, which beyond the study of the Forbidden City. It extended to a brand new field: imperial buildings, imperial people and technological modernization. It may look like the topics of the book are not related to each other, however, these three topics actually are created from one theme: the photographs are selected from the imperial family including architecture, people, military and economy, drawing the timeline from late Qing dynasty to Republic of China.

The majority of the selected photographs of *The Photographic Collection of the Palace Museum* are from the research library of the Palace Museum. In the preparation of the book, we receive huge support from the curator of the research library of the Palace Museum, Zhang Rong and Zhu Saihong; the librarians of the research library of the Palace Museum, Cao Li, Li Ying, Wu Anguo and Diao Meilin, provide us a reliable resource of photographs; the senior copy editors of the Palace Museum, Lin Jing and Zuo Yuanbo, improve the content of the book in terms of their academic research; meanwhile, Feng Yincong from the Forbidden City Publishing House, Liu Yang from the Management Department of Yuanming Yuan, Chen Dong from Chengde Municipal Administration of Cultural Heritage, Xiong Wei from Beijing Jiangong Architectural Design and Research Institute, and Han Liheng, Lu Wei provide their professional consultations to the book. We really appreciate their hard work and contributions to *The Photographic Collection of the Palace Museum.*

A 90-year's Palace Museum and 600-year's Forbidden City demonstrate a magnificent picture of Chinese history. While we are watching these old photos in order to study and learn the past, perhaps after hundreds of years, our descendants also try to unfold the history of our time through photos. We hope that the people who will be captured by cameras will illustrate a fantastic story to our descendants.

Editor

August 2014

图书在版编目（CIP）数据

故宫藏影：西洋镜里的皇家建筑/单霁翔主编. --北京：故宫出版社，2014.8 （2021.1重印）
ISBN 978-7-5134-0634-5

Ⅰ. ①故… Ⅱ. ①单… Ⅲ. ①故宫－北京市－摄影集 ②故宫－建筑艺术－北京市－摄影集 Ⅳ. ①K928.74-64 ②TU-092.48

中国版本图书馆CIP数据核字（2014）第163566号

故宫藏影

西洋镜里的皇家建筑

主　　编：单霁翔
副 主 编：赵国英　周苏琴
撰　　稿：王志伟
顾　　问：张　荣　朱赛虹　冯印淙　林　京　左远波　杨新成
　　　　　刘　阳　陆　伟　陈　东　熊　炜　韩立恒
审　　校：陈连营
资料提供：故宫博物院图书馆

出 版 人：王亚民
责任编辑：江　英
英文翻译：刘瑞溪
装帧设计：赵　谦
出版发行：故宫出版社
地　　址：北京市东城区景山前街4号　邮编：100009
电　　话：010-85007808　010-85007816　传真：010-65129479
邮　　箱：ggcb@culturefc.cn
印刷制版：北京雅昌艺术印刷有限公司
开　　本：787毫米×1092毫米　1/8
印　　张：51
版　　次：2014年8月第1版
　　　　　2021年1月第6次印刷
印　　数：13,501～16,000册
书　　号：ISBN 978-7-5134-0634-5
定　　价：360.00元